a love story
from the
end of
wor

Praise for *Beasts of a Little Land*

**WINNER OF THE 2024 YASNAYA POLYANA AWARD
FINALIST FOR THE 2022 DAYTON LITERARY PEACE PRIZE
A BBC WORLD BOOK CLUB PICK
INTERNATIONAL BESTSELLER**

'A stunning achievement' *TLS*

'Assured and textured' *NEW YORK TIMES*

'Epic' *HARPER'S BAZAAR*, Best Book of 2021

'Extraordinary . . . Gorgeous prose and unforgettable characters
combine to make a literary masterpiece' *KIRKUS* starred review

'A dreamy, intense debut . . . The prose is ravishing'
PUBLISHERS WEEKLY

'A potent and immersive reading experience'
CHICAGO REVIEW OF BOOKS

Praise for *City of Night Birds*

'This story left me thinking about the ways we overcome
setbacks and redefine what truly matters'
REESE WITHERSPOON, December 2024 Book Club Pick

'Impressive' *SAN FRANCISCO CHRONICLE*

'Engrossing' *LOS ANGELES TIMES*

'Lush prose buttressed by vivid details' *VOGUE*, Best Book of 2024

'A mesmerizing page-turner' *TOWN & COUNTRY*

'Lyrical, cinematic writing kidnaps the senses . . . [and]
made this reader hope for another Juhea Kim novel,
and another' *WASHINGTON POST*

Also by Juhea Kim

Beasts of a Little Land
City of Night Birds

a love story from the end of the world

juhea kim

THE BOROUGH PRESS

The Borough Press
An imprint of HarperCollins*Publishers* Ltd
1 London Bridge Street
London SE1 9GF

www.harpercollins.co.uk

HarperCollins*Publishers*
Macken House, 39/40 Mayor Street Upper
Dublin 1, D01 C9W8, Ireland

First published by HarperCollins*Publishers* 2025
1

ISBN: 978-0-00-878580-2

TO THE SOULS WHO
DESERVED LIGHT

SPIRITI DI MOLTO VALORE

AND THOSE WHO SHED
TEARS FOR THEM

CONTENTS

BIODOME

April 13. Almost midnight. Through the worn twill curtains, a viscous light was lapping into the apartment like flowing amber. Park washed his face in the bathroom, took his meds, and sat down on the sofa with the remote. One click, and the artificial blue light of the TV mingled with the sodium-yellow atmosphere of the room. He flipped through the channels. Game shows. Contestants competing for money, for marriage. The women are showing off, swiveling their hips and winking at the camera, and then they're ranked by the amount of applause they receive. People awwing over tiny puppies. Slow, close-up shots of some new hybrid food, a rare delicacy. A man has to eat a glistening pile of meat and cheese until his face is streaming with sweat and an exhausted howl escapes from his mouth. Woops, applause, groans, laughter.

A pair of professors seated facing one another against a black backdrop. No audience, no clapping.

"There are generations of children who are growing up not

knowing the meaning of the word 'rain,' 'snow,' 'clouds,' who have never seen the sea or the sunset in their lives. Are you saying that this is not a moral crisis?" said the pundit on the right side, a literature professor at Seoul National University.

"You are making a Romanticist error in conflating nature with morality. Nature in itself is neither good nor evil. Likewise, technology that is used to shape or control nature is neither inherently moral nor immoral. Less than two centuries ago, our ancestors argued over the morality of the 'Iron Horse.' Would you argue that the subway is evil because it bores through the earth and shuttles around humans in a sealed underground passageway?" said the pundit on the left side, a philosophy professor at Korea University. "Technology is strictly a matter of utility, not ethics."

"I would be greatly presumptuous if I were single-handedly conflating nature and morality, as you put it. But take this for an example: You know that language shapes thoughts, ideas, and morals. Even you must acknowledge that in our Korean language, the word 'God' is literally 'Dear Sky.' Long before I had ever romanticized nature, as you argue, our people have used words of nature to describe what is good and sacred, for five thousand years. When people no longer know what a sky actually is—and many of the younger generations have never seen it—how would they have any consciousness of a God?"

"You have a false understanding of the nature of faith. God—at least in the Christian sense—defies representation or proof. If one needed the sky in order to recognize God, then that would not be genuine faith at all," the philosophy professor said smugly, then carefully delivered the ultimate insult. "You speak out of sentimentality."

"I'm not being sentimental. I'm being human," his opponent replied calmly.

And on and on they went. Park turned off the TV.

"This is a nice place," Park said. They were seated in the courtyard garden of an elegant restaurant in Gangnam, surrounded by trellises and arbors of roses, jasmine, and other dark and lush shrubs that Park didn't recognize. The waiters in waistcoats flitted about discreetly like moths, lighting candles one by one.

"How many of these have you done?" Jina asked.

"What?"

"You know, *matsun*."

"Excuse me?"

"You are thirty-five and unmarried so you must have gone on at least two dozen."

"What else do you know about me?"

"The matchmaker gave me the usual specs. Your height, your looks. You're a senior engineer at the Department of Environmental Protection and you graduated from Seoul National University near the top of your class. But you don't have much family money or the ability to finance a new apartment for us. You're the only son of a widowed mother, which is guaranteed to scare off squeamish women. Physically, you have weak lungs, a common enough condition for those born before the Bio. However, I heard you have an IQ in the top 0.1 percent. So naturally, we may be a match."

"You certainly speak your mind, don't you."

"Here's what I think. I'm twenty-eight and the last of three daughters, so my parents are absolutely dying to get rid of me. My two older sisters both married at twenty-six although I'm much prettier and more intelligent than they are. My parents think it's because of my personality. I don't care much about shutting my mouth to flatter some man who is more stupid than I am. Do you care if I smoke?" she said, already reaching for her cigarette case.

"Be my guest."

"I'm terribly bored by my parents' endless entreaties and machinations. I'm just ready to end it all and marry whomever they think is appropriate. I'm tired. Do you know what I mean?"

"I think I do."

She smiled.

"You know, it's funny. I feel like I can be honest with you. It's not often that I get that sense at one of these things. The matchmaker did say we were an uncommonly good match with our astrological signs."

"Western or traditional?"

She raised an eyebrow.

"Why, traditional of course. Why would I put any trust in Western astrology for something as important as marriage? That stuff is just a bunch of nonsense."

"They seem about equal in my esteem," Park said.

She broke into a peal of silvery laughter. "I'm pleased that you disagree with me. I was worried that you'd be one of those scrawny, weak engineer types who don't seem to have any opinions of their own. And please don't take this the wrong way, but your first impression wasn't too far from my expectation. But you know how to push back—I like that." She took a long drag from her cigarette. "I feel like we're going to get along quite well."

Park wished he could come up with some sarcastic remark, but he couldn't. As for himself, he did not yet know whether he liked or disliked Jina. She had a pale heart-shaped face and long, shiny hair falling halfway down her back. She was wearing a silk sheath in canary yellow that complemented her ivory and jet-black coloring. By focusing his eyes on her bare arms, he could almost smell the perfume on her wrists and elbows—floral and slightly musky. Yes, he supposed she was a beautiful woman. He just wasn't sure if he could ever have any feelings for her.

"You look very nice," Park said, at last. Jina's eyes started dancing; she was used to compliments and had been on the verge of impatience with him.

"Nice?" she purred.

"Your dress. It looks beautiful." Park rambled a bit in his embarrassment.

"Oh yes, I suppose. It's yellow so I figured I'd wear it. I see you're not wearing anything for Yellow Day? You don't have a tie or something?"

Park didn't have a yellow tie or socks or anything like that. All day at work, colleagues had teased him about not being in spirit. "Being a spoilsport, Manager Park? Surely you don't want bad luck in the next year?" One of the men, a junior engineer, even jokingly offered to trade a sock with him, so they can each have one that is yellow. But grateful though he was for their good-natured banter, Park was secretly glad about not participating in the whole thing.

"I don't really believe in that stuff," Park said.

"Goodness, you don't believe in anything. Not astrology, not Yellow Day . . ." Jina smiled. "Is there anything you do believe in?"

Indeed, what did he believe in? He did not know. He was only ever sure about the things in which he couldn't believe.

The morning after Yellow Day, New Seoul was cast in sepia as usual. There was once a time when trees turned gold in September and October, or so Park was taught in biology class; but the leaves in those photographs were nothing like the color of light here, which was a muggy reddish-brown like distant memory. Instead of autumn leaves, Yellow Day commemorated the Yellow Sand that blew in from the deserts of China and Mongolia every spring, carried by the west wind. This was a phenomenon that was documented in the earliest annals in ancient Korean history, going back at least two millennia. Through most of the twentieth century, the sandstorm happened three days a year in April, leaving a thick layer of dust over everything in the whole country, city, and countryside alike. Then as the desert in China grew ever larger, the sandstorm lasted longer every year: seven days, then twelve, twenty-five, forty-three, sixty-seven. Every consecutive year, the west wind carried even more sand over the Yellow Sea. By the time Park was born, it was calculated that the sand dumped on the Korean peninsula was one million tons a year—enough to fill 66,667 dumpster trucks.

One of the earliest consequences of the Yellow Storm was that there was no longer any spring. No more flowers or sitting outside on a park bench dazed by the scent of the sun warming up the grass. During the day, the sky was always dark gray and thick with particles; at night, it glowed an unearthly red. People couldn't go outside during the Yellow Storm: as soon as you opened the front door, the sand would sting your face and bare skin like a thousand bees, and swarm into your eyes, nose, and mouth until your lashes crusted over with ash. Even after the storm receded, the toxic sand laced with heavy metals and pesticides remained in the atmosphere so that every breath one took, inside or outside, could cause disease—asthma, interminable coughs, rashes, cancers, and blindness. The sand mingled with clouds in the atmosphere and came down as acid rain, melting down trees and crops. Food prices, already high, became astronomical. There was nothing left on supermarket shelves, even if you could afford it; and anyone who could afford it left the country. Reservoirs and water supplies were found to be contaminated, so no one could drink or even cook with tap water. The rivers became thick with thousands of dead fish. There were protests demanding government action, and eventually riots—

eruptions of chaos so that the scant resources could change hands before disappearing completely.

In New Seoul, the poor were counting every grain of rice before a meal, which started from one hundred and dwindled to just ten pieces for dinner. When even that ran out, a family would curl up in bed together holding hands, knowing that they would not wake up.

They were saved at last, and only, by the Biodome—completed on April 14, thirty years ago. The first Yellow Day. When it was built, no one objected to it saying things like the sanctity of the sky or the humanity of nature. They were all glad to be alive and to breathe the air without worrying that it would cause cancer of the throat or lungs. The Bio (BEE-oh), as it was also called, was a clear enclosure of fifty-kilometer diameter over New Seoul, blocking off the Yellow Sand and letting in what remnants of sunlight could penetrate the particles. The Bio's internal atmosphere was always a foggy sepia—not because of the sand passing over the enclosure, but because the combined glare of millions of neon lights reflected back on the inner surface of the Bio as a volcanic, red-tinged brown, every day and every night.

Naturally, the Bio also eliminated all other precipitation along with the sand, but they found that stable internal humidity and underground watering systems eliminated any real need for rain. It was truly an engineering feat, the world's first successful habitable indoor enclosure at such a scale. And this turned out to be a case in which the problem also provided the solution: Bio technology became Korea's most important and lucrative export, so that even while importing nearly all food and other products that they no longer made themselves, the country as a whole became more prosperous than ever.

People soon forgot that they were living under a dome. It wasn't that they forgot it existed, since the Bio was now their livelihood as well as their lifeline. They just ceased to remember that they were under it. After all, it wasn't visible or tangible for most people living inside, and life was better again. Once something became a part of the environment, people accepted it without question and, furthermore, forgot that things had ever been otherwise. It took about a month for the very intractable, but for most it was a matter of a few days before they could no longer abide by even the memory of open space. So Park found the television special the other night strange, with the two aging intellectuals arguing vociferously against each other. And also rather sad, Park thought, to observe those scrawny and balding professors sitting in their ill-fitting suits, each with a glass of tepid water in front of him, discussing what no one thought about with any seriousness or urgency. It had to have been for the thirtieth anniversary celebration. Why that kind of public debate was even included alongside the usual fluff, cooking shows featuring yellow ingredients or dating contests, he didn't know.

Park himself thought about the Bio quite a lot, but only at his job: his duty as a senior engineer in the Department of Environmental Protection, Office of Human Conservation, the Biodome Management Bureau, called for constant monitoring of the main Bio (now there were four total in Korea, and twelve overseas). He had been five when the Bio was built, so he had no true memories of life before it. Sometimes though, he thought he could remember being on the beach, the heat and the brightness of the sun on his skin and the cool wind that shook his hair, and the achingly vivid blue of the sea. He had had dreams where he felt so sure of the sea breeze caressing his cheek that he woke up laughing. But his mother assured him that he'd never been to the beach in his life. By the time he was born, the Yellow Storm had forbidden all but most necessary excursions outside. His mother said he had never gone out of the house until after the Bio.

There was no wind inside the Bio, only the mechanical exhale of vast ventilators creating corridors of fresh oxygen through the interior smog. So how could he imagine that pressure against his cheeks, the whipping of his hair? How could he imagine something so real if he'd never experienced it? The same way it feels so natural to fly in a dream, Park told himself. After he determined that, he stopped feeling the sea breeze in his sleep. The whole thing put him down with a leaden sense of loss, if it's possible to lose something one has never had.

It was better not to think about the things that weren't here and now, and yet Park had this tendency to drift—and when he consciously stopped himself, he felt caught between two worlds, one occupied by everyone else and the other where he was the sole citizen aside from his strangely companionable thoughts. That latter world wasn't real; so he did his best to cling to the first, blending in with the others with his quiet, unassuming demeanor. His unimpressive appearance helped him move along unnoticed: his bony frame with small shoulders, hollow chest, and unobtrusive, mediocre face gave the impression of a smart though unthreatening engineer-bureaucrat. His teachers had all but pushed him along to become exactly that, praising his even temperament and cognition without rebelliousness. Despite malnutrition from gestation to age five, the most formative years for his brain, he had excelled in his exams and personality tests as required. Bioengineering was the most promising field to which the brightest students aspired, so he'd applied to that department in the best university in the country and had been accepted. After graduation, he'd applied to just one job at the Department of Environmental Protection and had stayed there ever since.

In short, the nature of Park's effortless existence was akin to being pushed by the crowd on a packed subway platform. During rush hour, he didn't really walk so much as let himself be picked up and carried by the compressed mass of bodies around him, moving in mindless unity like a school of fish. Getting out at his stop at Yeouido required much more maneuvering, and Park carefully picked his way among commuters until at last he stood at the foot of his building. It was already 8 a.m. when he arrived at his office on the sixty-third floor. His colleagues greeted him with inquiries about the previous evening's date. Despite the fact that Park would never volunteer such information, they had figured it out with just one look at his best suit. Harassing unmarried coworkers over their blind dates was as much a time-honored tradition as the *matsun* itself.

"So how was it?"

"Fine, I guess."

"What, just fine? We all looked her up online. She's gorgeous."

"Her profile says her hobbies are 'playing the piano' and 'fencing.' What a catch!"

"If you don't want her, I'll take her off your hands."

Park smiled vaguely and sat down at his desk until they went on gossiping on their own. He set down his briefcase, took his meds—the first two pills of the day—and turned on his screen. His first task was always checking the full report of the vitals of the Bio. The oxygen levels were sagging, which needed adjustment—but elsewhere he found that trace gases were unusually and concerningly high. He spent the next hour pulling up more data, and then rushed to talk to his supervisor, the chief engineer of the Bureau.

In due silence the chief engineer scrutinized the tiny green numbers filling up the black screen. He kept inhaling sharply and audibly through his nose, which appeared necessary to maintain his train of thought when it came to the most serious calculations.

"This amount of carbon monoxide isn't something to be so alarmed about," he said at last.

"But the spike in sulfuric acid?" Park asked hesitantly.

"That's likely from Yellow Day celebrations."

"It's at its highest level in thirty years."

"So what are you thinking, Manager Park?" The chief engineer gazed at him steadily.

"I think there is only one plausible explanation. The sand is getting inside the Bio."

"But the air pressure is normal. It's probable that the sulfuric acid number is a fluke. False positive. Based especially on the fact that the number is so high. The Bio contains twenty thousand cubic kilometers of atmosphere, carefully calibrated to support life. Even if there was a crack or a fissure somewhere, the amount of sand that can get in wouldn't make any noticeable difference in the air composition for weeks down the road. I thought you were a smart guy," said the chief engineer. "Enough of this. I have other matters to attend to." Then he turned to his screen, effectively ending their conversation.

Park left the office feeling agitated and almost indignant, but the chief engineer's argument was hard to refute. Still, he had a hunch that this wasn't some false positive. When he saw the data, he'd immediately gotten chills down his back. At the end of the day, the Bio was man-made, liable to break down just like anything else.

Over the next week, Park kept a careful watch on the data. After that initial spike, the sulfuric acid levels stayed constant during that time, which appeared to support the chief engineer's hypothesis rather than his. Park felt a strange gaping hole in his chest and became unsettled by the realization that he was incredibly disappointed. It was not because of his ego—he didn't care that someone else was right over him. It also wasn't because he got a twisted, psychotic thrill over the possibility of chaos, destruction, and tragedy. He could only describe it as similar to something being taken from him, a sense of being robbed, though that notion was utterly absurd. Robbed of what, exactly?

Nevertheless, his sense of not-having was so real that he found himself calling Jina for the first time since their date. And yet, their exchanges of greeting were so awkward on both sides that he almost wished he could hang up.

Instead, he carried on.

"I was wondering if you wanted to meet," he managed to say.

"For what?"

"I thought that you might like going to a recital." He had looked into music listings and found a recital of a celebrated pianist that weekend. Jina softened as she listened; then they agreed on a time and place to meet and hung up before either of them could begin to feel awkward again.

On Saturday, Park waited for Jina by the fountain in front of the performing arts center. He spotted her slim figure across the plaza and waved stiffly. She was very dressed up in a mercury-colored evening gown, and Park immediately became self-conscious that he'd repeated the same best suit he'd worn to their first date. She slowed down as she approached him, as if she was tempted by the possibility of turning around and leaving. Park reddened; the sound of falling water filled his ears. Jina took a deep breath and finally reached near enough for them to talk. But Park hesitated: they didn't know each other well enough to use the informal speech, speaking aloud her name was overly intimate, and the formal *hello* or *how are you* seemed fatally unromantic, even to him. With great embarrassment, he settled on, "Let's go inside." She seemed to accept this as a greeting, and allowed Park to rush to hold the door open for her.

They found their seats in the auditorium. The lights slowly darkened, and the audience broke into applause as the pianist was led onto the stage by a young woman. They stopped next to the piano, and he bowed, to more applause; then she pulled out the bench for him and he sat down. The pianist was renowned not only for his virtuosity, but also because he was blind. He'd already been an international concert pianist when he lost his vision, just days before the Bio was finished. When asked in an interview how he adjusted to playing without sight, he had said that he was simply doing what he'd always done, seeing the music and the piano keys in his mind. In front of a piano, it was as though he'd never lost his sight.

And as he started playing Beethoven's sonatas, Park was struck by the impression that the pianist really *could* see: not just the piano, but also something beyond their reality, the concert hall, their anthracite forest of supertowers and underground malls and apartments like catacombs, every habitable space filled with people or something to support them. No one else objected to this teemingness as far as Park could perceive, but it was all too much and not enough that he sometimes couldn't even stare down his bowl of salad. The pianist wasn't playing for any of that, Park realized—he was playing for that unknowable beyond only he could access. Without meaning to, Park's eyes became hot with tears. For the first time in his life, he felt as though someone had looked into his eyes and said to him, I know what you yearn for.

He slowed down his breathing so that Jina wouldn't catch him crying. She turned to glance at him a few times but mostly kept her eyes on the pianist, her chin slightly lifted, her posture elegant and proud. Park could tell that she knew all the pieces by the way she seemed to anticipate every change in the theme, and how she smiled slightly before certain passages she liked. That she also loved music gave him hope that they may truly become a couple. He had never had a real relationship before. All of his previous setups by his mother, matchmakers, and online dating companies had fizzled out after a few dates. He had never disliked any of the women, who were more or less similar: pale, feminine, soft, and a bit anodyne, like cotton candy. The problem was that neither Park nor any of the women could ever muster enough interest to keep going after the first two dates.

Park still didn't know how he felt about Jina. Underneath her delicate physique, she had something sharp in her. She smoked, and she flirted without giving any indication that she genuinely liked him. She understood music. Park took a sideways glance at her profile and noticed, with pleasure, how she'd made up her eyelashes dramatically with mascara. She was wearing a long silver dress with a low-cut back, which was very flattering to her figure. The fact that he took note of her details seemed to be a good sign.

After the recital, they followed the crowd out of the concert hall and onto the plaza. It was another muggy, red-tinged night, but Park wanted to linger there for a moment. The arts center sat on a piney hill in the middle of New Seoul, so it had a view overlooking the entire north side of the city. He led Jina to the edge of the plaza, where they leaned out over the balustrade.

"What did you think of the recital?" he asked.

"I thought he was brilliant and sensitive in his own way but also not powerful enough. His style of playing is really much better suited to Chopin," she said. "You must think I'm a snob—it's just that, being a pianist myself, I have different standards than nonmusicians do. I also have a more definite idea of how these pieces should be played—having played them myself and heard so many recordings—that it's hard to overcome certain prejudices."

Park thought she was perfectly justified, and yet he no longer was in the mood to talk about what the music had made him feel. Instead, they both looked out over the city in silence: the super-towers rising nearly to the top of the Bio, and vertical forests and gardens reaching hundreds of stories, all connected by a web of skywalks, monorails, and elevators, and on the ground level the tiny moving figures of electric cars. Everywhere, thousands of neons flashing and pulsating like colorful drumbeats. There were so many lights crowded together in this city that there was never any complete darkness unless you covered your face with your hands. Or, unless you were blind. That suddenly struck Park as very strange.

"Where do you live?" Jina asked. "Can you see your building?"

Park's building was hidden by the others, but he could point out the general direction of its location.

"Over there, that's my building." Jina pointed to one of the towers. "I'm on the hundred and ninth floor. Right about there, where my finger is. What about you?"

"I'm on the fifth floor."

"Goodness, you are practically underground," she said reflexively, sounding disappointed. The best and most expensive apartments were on around a hundred stories, since that was the level of most skywalks and monorails. Jina had never lived below the ninetieth floor in her life, and she wasn't about to start now. Every passing minute she was more convinced that any intrigue she'd initially felt about this man was her own ennui acting out in final desperation. When the matchmaker had presented his profile, her parents had in fact been cautious and unimpressed. Too poor and unconnected, they'd said to her. As an individual he may be your equal, given his educational pedigree, but his family is completely below our family. Still, she'd found something touching about his photo, liked his serious, intelligent eyes under the charcoal black eyebrows; and he'd written that he was a fan of classical music. But when it came down to it, he was like every other blind date she'd had, only smarter and poorer. She glanced at him out of the corner of her eyes and was freshly annoyed by his thin, slightly concave chest and his spoonlike profile, characteristic of those born before the Bio. As for supposedly being a fan of classical music, he had absolutely nothing to say about the performance, and was just standing there like a statue staring out at the violent lights. So he was one of those boorish men who claimed to know something about music to flesh out their profiles. Perhaps what had most appealed to her was the chance to frustrate her parents by falling in love with someone who they thought was beneath her. But such capriciousness was costing her her youth. After the night was over, Jina vowed to have an honest conversation with her parents and marry the next eligible match, a younger and richer man who could afford a new apartment in the right district, nowhere near the ground.

As for Park, he was thinking of what one of his colleagues—the junior engineer who offered him one yellow sock—had told a group of them over lunch about his new girlfriend. The guy had said that he and his date were on the monorail when she pointed out her tower, her apartment just a tiny yellow square somewhere in the middle. Every time he passed by it now, he looked at the cluster of windows knowing one of them had her in it, and the awareness made him unbearably happy—and that was how he knew he was in love. Where someone lives is the most uninteresting, boring fact—unless it's someone you love, and then it becomes the most imperative thing to know, he claimed sagely. Think about it—he said—do you all remember where I live? I've told you guys a million times. When you run into someone from university and you ask which neighborhood, which building, simply to be polite—do you actually remember it? You know you have no recollection whatsoever after a day or two. They'd all broken down laughing.

Park had zero interest in knowing where Jina lived.

He looked at her, clad in her thin silvery dress, staring out vacantly into the vast expanse of lights. Nothing could bridge the distance between them, not even standing here together at the edge. He thought about how he would never call her and she'd never call him, and how they would forever disappear from each other's lives.

It was close to midnight by the time he came home. Park washed his face and took his meds—the seventh and eighth pills, the last dosage of the day. The meds had kept him from coughing up bloody phlegm, made him weak but alive, every day of his life as long as he could remember. Park felt weary, but he still couldn't fall asleep. He turned on the TV and flipped through the channels, and stopped when he found the rerun of the Yellow Day public debate program.

"Your argument is based on the assumption that technology doesn't concern morality. But the real moral issue of the Bio is that it eliminates choice," said the literature professor.

"What nonsense—people are free to leave the Bio whenever they like. People travel and go abroad," said the philosopher, smirking. "I've myself just come back from a visiting professorship at the University of ___."

"Then you know well that only the wealthy can travel. How can an average person afford the Bio reentry fee and the airfare?

Effectively, no normal citizen steps outside the Bio in his or her lifetime."

"People always have a choice. It's not necessarily a choice between option A or option B, like at a restaurant. Sometimes you are given just one option, A—but you still have the choice to refuse it," the philosopher said. "In this case, anyone who may wish to do so for whatever reason can leave the Bio. There is no Bio exit fee, you might remember, or any law that mandates that a citizen stay within a Bio. If you so desire to leave, that is perfectly within your rights, although doing so is clearly against your self-interest. Even a starving person may refuse food. Or one may even refuse to keep breathing, if they so choose. They have free will and choice at every moment in their life, no matter what circumstances they are in."

"But that is exactly the thing against which I am arguing. A starving person who is offered food doesn't truly have a choice between survival and death. They can only exercise free will when they have a choice between viable options," said the literature professor.

Park got up from the couch. He rushed to get his notepad and pen, and wrote this down:

> *Morality begins with choice. Without choice there can be no good or evil. There is no freedom here. Without freedom, there can be no meaning.*
>
> *Without freedom, there can be no love.*

When he finished, he read and reread what he wrote. The sense of being wakened to a new consciousness terrified and exhilarated him at the same time. On the one hand he was filled with tremendous relief at knowing, at last, why it was that he felt so disinterested in his own life. He had never managed to buy into the things that others so naturally believed were important—money, superstitions, marriage, copulation, apartments, and all the ways in which people proliferated under the Bio like microbes in a Petri dish, meaningless and abundant. On the other hand, he still couldn't decide what he should do. Should he leave? Could he? He understood nothing about life outside the Bio, or had any knowledge or skill besides the Bio itself. But that also gave him an enormous advantage: he was one of the few people in the entire city who knew the maintenance exits used periodically by the Department crew—underground passageways that lead out a hundred kilometers to the east, where the mountain range blocks some of the sand. Some communities remained there, nestled in the deepest valleys. As recently as several years ago, Park had read some article or another about a handful of farmers who had managed to raise potatoes in the gullies between the mountains and the sea. No one in the Bio remembered them, so they took on the ghostly shape of hearsay, but without news of any kind Park had to assume that they still lived.

Perhaps in the morning his resolution would fail, his lungs would collapse, and he would resign himself to obeying the Bio's rules without struggle. There was a chance that some hideous punishment was already being prepared for his insubordination. But somewhere he could hear and was soothed by the arpeggio notes of a deaf composer. There was only one thing he could do in the here and the now. With a smile, he closed his eyes and waited to be swept away by the wind.

COLOR OF THE NEW WORLD

*En vérité, ce qui donne à l'objet sa couleur, ce n'est
ni ce que l'on nomme la couleur réelle ni la couleur
conventionnelle . . . C'est la vie qui crée les contrastes
sans lesquels l'art serait inimaginable et incomplet.*

It is neither the so-called real color, nor
the conventional color that truly colors the
object. . . . Life itself creates contrasts, without
which art is unimaginable and incomplete.

—MARC CHAGALL

It is a blue and white day on the Chartreuse Mountain. Between a pair of birch trees draped with moss and lichen, she touches snow for the first time in three years, unlacing her hiking boots and carefully stepping onto the pure patch. Her toes sink into the powder and grip the mossy rock beneath it, and she steadies herself, placing one hand on each tree. She considers with pride how far she's traveled to be standing here, sharing the roots of these trees and drinking in the snow of the Alps. It has taken her six hours of flight from New York to London, another two hours to Lyon, another hour and a half bus ride to Grenoble, then four hours of hiking in the mud to see this. To touch the purest thing created by the sky. She laughs, sniffles. Small tears leak out of the corners of her eyes.

From beneath her feet, the Chartreuse range rolls out and joins with the Vercors on the right and the Belledonne on the left. A pale sun looms over their blue peaks, hatted with snow; and the wind sifts the meringue-like brume in the basin between the mountains. The vertiginous silence goes through her jacket and pricks her in the ribs. It's almost painful, this happiness, hovering just a shade above loneliness. She holds an imaginary brush in her right hand and paints the sky.

But the snow disappears quickly on the way down, and the trail manages to be both boggy and jagged; she has to pause several times to kick off the bowl-size mud crusting on her boots. She had heard that even the sides of Mont Blanc are running brown with melted glaciers. When the highest peak in Europe runs out of snow, where and how many hours will she have to fly to see winter?

As she wonders, a flame-colored fox jumps off a cliff and dashes right in front of her onto the trail. It's surprisingly huge— the size of a broad husky, with twice as much tail. Before she can cry out, it turns around and leaps away.

When she's collected herself and begun walking again, she hears a noise from behind her and thinks the animal might have returned. Instead, a cross-country runner sprints recklessly down the muddy, rocky trail in her direction. She steps to the side and he gives her a nod in acknowledgment, disappearing like a train skipping an unimportant station. Then, because she's just had a *moment*, and because she hasn't talked to anyone in almost a whole day, she shouts to the back of his head: "*J'ai vu un renard là-bas!*"

At first she thinks he's going to keep running, but he slows down and stops, breathing hard. She thinks he throws her an accusatory look over his shoulder; he hadn't wanted to be disturbed. "*Pardon?*" he growls hoarsely.

She instantly regrets her overture, but it is too late to take it back. "I saw a fox just now. It ran down from that cliff, turned around and disappeared that way," she says haltingly in French, trudging toward him.

"Ah, they live around here. I see them sometimes."

The runner trains his eyes on her face, and she wishes she hadn't washed away her mascara with tears of joy. "It was much bigger than an American fox," she stammers.

"That's funny. Everything's bigger in America except for foxes, apparently. And you weren't scared?" He props his hands on his hips, shifting his weight slightly to one side. Toward her, it seems.

"*Pas du tout.* Not at all."

"You're not scared hiking alone?" he asks again, insistently.

"No. Not at all. Are *you* scared, running alone?"

He laughs, crinkling both his green eyes and his boyishly upturned nose. He's wearing a pair of those black performance leggings that make most men look vulnerable or confused, like they've pulled on their girlfriend's yoga pants by accident. On him, the spandex is a supple armor, stretching over a body accustomed to either agility or relaxation and nothing in between. All nonchalant, Spartan grace, she thinks. If she hadn't stopped him, that body would have kept running forever.

"No, I'm not scared either. *On se ressemble, vous et moi,*" he says, extending a hand. "*Benoit, enchanté.*"

"*Enchantée.*" She accepts the warm hand. Immediately, she thinks, *why here, how now, what next, where to.*

She measures out the days in Grenoble not by coffee spoons but by cups of herbal tea. Each time she goes over to his place, that's what he serves her. Benoit isn't like most Frenchmen. He doesn't drink coffee or wine. He doesn't smoke. He's a professional climber, so with it comes all the training and the healthy lifestyle. So after they have sex, after she licks his perfect abs and squeezes her thighs around his taut hips, they have herbal tea in his kitchen. She begins to associate the flavor of lavender and licorice with making love.

Benoit hands her a mug and sits across from her, leaning back in his chair. He asks her questions, and she tries her best to not mix up her *masculins* and *feminins.*

"Why did you leave New York? And what are you doing in France?"

She wonders how she can describe the changes in her city. When she first moved there, she had rented a one-bedroom apartment for $800 a month in Bushwick. New York is a harder place to live now than before: the tropical nights, the millionaire condos over rat-infested subways, the art crowd gnawing on the bones thrown by the Guggenheims and the Gagosians. There was the circle of self-actualizing, coke-loving, and costume-wearing Burners, of which her Brooklyn friends were a part. There was the circle of cool, young curators who had done their PhDs at Harvard or Penn and were now working at MoMA, like her friends from university. There was an art-tech contingent. For example, a guy had gotten funding from one of the Gs and created some startup that now employed hundreds of people in an entire glass-walled building in SoHo. It was changing the world, according to the founder, whom she dated once upon a time. These groups gathered and dispersed across town, mesmerized by NFT videos (hideous and worth millions in crypto) or silver body paint or EDM or crystal dildos or scent-art journeys, and she didn't understand any of it. Didn't like any of it.

And then the seasons changed. First the summers became unbearably long and hot, stretching from April to October and then March to November. She slept completely naked over her sheets while the temperature never dipped below ninety degrees for twelve straight weeks. She felt drained waiting for autumn to begin, but the bright foliage never came; the leaves went straight from green to a mothy brown.

Then, two years ago, she realized a whole winter had passed without a single snowfall. No one else remarked on it. Her friends looked strangely at her while they dined al fresco in shorts on Valentine's Day. Why was she complaining? they asked. It was nice not having to wear a coat.

Every morning, she woke up and wondered whether it was worth making another painting, given the circumstances. Before deciding to leave, she hadn't worked on anything new in over three months.

It's all too hard to explain in French, so she gives another answer.

"I realized that in New York, I was never going to fall in love again."

This is how she spends her month in the mountains: some evenings, she goes over to Benoit's house and he makes her dinner. Sometimes she cooks instead, and he wraps his arms around her from behind in a way that makes her feel adored, if not quite loved. While watching a movie, he massages her bare back in a new way, catching the flesh between his forefinger and thumb and slowly rolling the pinch up her spine like the world's tiniest snowplow. (Years from now, it is this insignificant detail she will remember most clearly, since only minor pieces are retained from minor relationships. But whether such memories can leave a permanent mark on one's psyche is determined not by depth, but by surprise.) She helps him stretch his hip flexors. He takes her on a night hike with headlamps and they gaze down on Grenoble, a lake of lights beneath their feet. She introduces him to American and British rock of more recent vintage than Pink Floyd. The fact that he approves of her musical taste excites her. They listen to her playlists, one earbud apiece, bundled under a Marseillaise quilt in front of an ancient fireplace.

"So what are you going to do after Grenoble?" he asks when the music stops. His voice is so casual—borderline cheerful—that he could be saying this to an elderly visiting relative. She takes careful note of his lack of anxiety.

"I'm going to the Côte d'Azur. Nice, Antibes."

"And then after that?"

She has been thinking about this herself, wondering whether she will return to New York. She has already imagined, dozens of times, living in this country house with its French-blue shutters and spawning beautiful half-French children. But she'd rather run a stake through her heart than let him know how often this scene has played inside her brain.

"I guess I'll go back home. Will you come visit me?"

The way he looks at her makes her regret wanting to extend whatever it is they're having. Once, a critic from an online art journal wrote that she wills her paintings to go beyond the canvas. At the time, the comment had satisfied and flattered her. Now she thinks that it's her fatal weakness. She is always confusing the boundaries between what's real and what's not; always crossing the line and then getting broken, never mind the lessons from the past.

She tries to hide her disappointment behind her mug of tisane, but they're both aware of what's going on. At the midpoint of his thirties, Benoit considers himself not unfeeling; he's been in love before, he's been hurt before. He grows tender as he takes in her hopeful brown eyes, high cheeks strewn with freckles, the shadowy hair that falls across her face. From the first time they met, what has struck him most about her is how unafraid she is of anything. Even now, when her whole face is red from nerves, she actually seems braver than he is when he's free-climbing on Mont Blanc. Although he hasn't seen a single painting of hers, he is certain she is an artist from the way she sees the world, her amazement at life. He is amazed by *her*.

But is it love when it is so finite? He isn't sure.

"You're so, so beautiful, *ma jolie américaine*," he says, instead of empty promises to visit her or stay in touch. This is honest. He takes her hand, plants a kiss on her smiling lips.

The Cape of Nice is like a white sailboat moored in a sea of memories. She stands at the prow, her hands falling open to the sun-scented breeze. Her eyes are full of bright rocks and deep emerald waves.

She strips down to her bathing suit and picks her way to a cove. There is an iron ladder fixed to the side of the cliff, its last step hovering about four feet above the water. She dangles for a moment, and then lets go—disappearing into the sea, leaving in her wake a circle of shimmer like spilled sugar.

After swimming around for a while, she comes back up the tangle of rocks. Someone else has arrived there, she notes with displeasure. It's a man working on a laptop, which feels insulting to the scenery. From his expensive shirt, aggressively Californian jeans, and fake-spiritual bracelet, she decides he looks like an American.

Back in the water, she thinks about Benoit. He never did anything to maintain the illusion that this wasn't a permanent farewell, other than almost reluctantly trading Instagram accounts on the very last night. It crushes her, not just his indifference but also the unfairness of the whole affair, like exchanging dollars for euros and seething at the reduced sum. But of course, the money changer asks, that was the deal to which she agreed, wasn't it?

She reminisces with near reckless abandon, because she knows that the sea will exorcise him with or without her volition. In February the water is cold, so she can't really hold any thoughts other than keeping up her body temperature. Love is too complicated to contemplate, and in due time it is physically overcome. It seeps out of her and rejoins the sea.

The American with the fake-spiritual bracelet is there when she climbs back up, dripping in ever-paler shades of green water. She has wondered whether the intense cobalt of the cape could be retained in smaller and smaller amounts, until the drops left on her body sparkle with the tiniest particles of green. The American takes a sip from his coconut water without lifting his eyes from the laptop. The whole time she is toweling off and getting dressed, he doesn't glance once in her direction. He forces her to leave by not acknowledging she exists.

When the church bells ring at nine o'clock, she wanders down to the flower market in Cours Saleya. The vendors arrange into perfect pyramids the glistening strawberries and the tightly wound endives like sealed letters. The gnarly, huge lemons from Corsica tumble around in wooden crates; anemones and ranunculuses are beaming next to topiary kumquat trees.

She pauses in front of a young African who has spread a piece of fabric on the slight emptiness between the florist and the fishmonger. He meets her eyes and emits a soft *bonjour*, which she returns even more quietly. Senegalese, she assumes by his intonation. His skin is Van Dyke brown, dark enough to absorb light. Sitting cross-legged, he is carving animal figurines out of a piece of wood. A menagerie of elephants, zebras, lions, and rhinos quietly stands guard around him, marking a tablecloth-size continent. Wood chips fall away and roll around his feet like popcorn. She considers getting a piece but walks away because she has once read somewhere that Westerners buying these figurines is causing a massive destruction of a rare and precious African tree.

A few blocks from the market, she walks straight into a miraculous little art-supply store. The sign outside in flowing cursive claims that the shop was founded in 1850. In a glass cabinet filled with vintage paint tubes, she discovers a real Sennelier emerald-green watercolor from the 1920s. It's still usable, the shopkeeper assures her.

The next morning, she sets up her easel on the white rock and carefully squeezes the Sennelier onto her travel-size pan. The coil of paint is surprisingly rich and supple out of the century-old tube, like a small lizard. On paper it becomes translucent and reflective, as though she is painting with seawater. She thinks, *careful, careful*, even as she falls under its spell.

Her hand moves, following the directions of her eyes, bypassing her mind. When it's finished, she stares at it, dazed. She's never painted anything like it before.

"How much?" asks a voice from behind her, in English.

She turns around and sees that it's the American with the fake-spiritual bracelet. He is gazing at her watercolor, his arms crossed over his chest.

"It's not for sale. It's for me," she answers curtly in French. She has just realized he reminds her of the art-startup ex, down to the knuckle-size silver ring that surely is from a life-changing trip to South America.

"Ah, I didn't mean to bother you." He switches to a perfectly bourgeois French. "I really love this work. The composition, the energy. And this color is so . . . I can't put a finger on it. What is it called?"

"It's called emerald green," she says, unable to resist the chance to show off a little in front of the snobby man. "It was discovered in the early nineteenth century and was a favorite of all the Impressionists. Picasso used it during his Blue Period. But it's been banned since the mid-twentieth century."

He frowns. "Banned?"

"It's very beautiful and very lethal. A mixture of copper and arsenic. Napoleon is said to have died from poisoning by his emerald-green wallpaper."

He winces slightly, gesturing at her painting, and she enjoys his discomfort. "So what is this?" he asks.

"Well, nowadays you can find synthetic emerald green, or use cobalt or viridian. This, however . . ." She smiles, her finger hovering above the paint coil in her pan. "It's the real thing."

"I'm now glad that the painting is not for sale." He smiles. It's an objectively good-looking face, but somehow the expression hides rather than reveals who he is. She is familiar with that smile, knows how to play the game.

"After it's dry, it can't really harm you unless you have a thing for licking your art," she tells him. "If you still want it after all I've told you, it's yours. A gift."

She is surprised by her friendliness to someone she has secretly despised for close to a week. It's just that she's trying to fill a void, she reasons. She needs distractions. He isn't what she wants, but at least she wouldn't be alone.

Still smiling, he says, *"Volontiers"*—*I'll take it.*

His name is Leo and he's not an American, as she had guessed, but French-German—originally from Strasbourg. After Sciences Po, he got his master's in engineering at Stanford and worked for Google before founding his own startup. This company is doing so well that it is no longer accurate to call it a startup; it has offices in both Americas, Asia, and all over Europe. It's headquartered in Sophia-Antipolis, the so-called French Silicon Valley just outside of Nice. All this he explains over dinner, where candlelight dances over the black-polished mirrors and the giant marble urn of mimosa flowers in bloom.

"So what is it that your company does?" she asks, setting down her glass of champagne. "Changing the world, right?"

"Ah, you're making fun. But yes, we are changing the world. Tell me, what is it that Google did that no one else had ever done before, in the history of humanity?"

"Changed the way we find information?"

"Very good. But it's more than that. It's changed how we get information, but also what learning means—even how our brain thinks."

"And your company is going to be the next Google."

He leans back into the banquette and smiles. "Better than

Google. I said they changed the human brain. My company is going to change the human consciousness."

"Aren't they the same?"

"No, of course not. You didn't paint with your brain, did you?"

She blushes and concedes defeat. "So how are you going to change human consciousness, exactly?"

"We're working on a project. A new invention," he says, and she isn't sure whether it's just the candles or something else that's making his eyes glow so suddenly. This is the first expression that reveals something about him.

"Listen, *Leo*." She stabs her fork into the bowl of linguini and laughs without smiling. "What I wish for is technology that solves real problems, not some new way to order food or share selfies with strangers. Don't you see? The world is being destroyed as we sit here in this chic restaurant drinking champagne."

"In what sense?"

"Have you read any papers? Seen the news? California, which used to be as beautiful as the Côte d'Azur, is burned to a crisp. They're saying we will have an ice-free Arctic Ocean in the next year. There's the giant garbage patch the size of China, floating in the middle of the Pacific. What about the millions of refugees from the famine and drought in Africa and the Middle East getting stuck in border camps or drowning in the Mediterranean? And it's not like Europe is free from harm—last summer, Berlin was a hundred and twenty degrees, as hot as Baghdad."

"These problems have always existed throughout time," he says, taking another sip. "All you can do is play your part and wish for the best."

"It's too late for that now. Say we manage somehow to survive as a species. There won't be any beauty in it. I would like to see it all before the world ends . . . that's why I'm here."

"As for beauty, that's a matter of opinion. But the world won't end." Leo says this with less flippancy and more assurance than she expected; he almost sounds like he could be hiding a kernel of sincerity under all the designer layers. He motions the server for the check. "Technology will advance exponentially in the next few years. We're about to see a massive economic and societal disruption—the Fourth Industrial Revolution. There will be winners and losers, just like always. But this is not the end. Come," he adds, rising. "I'll show you something."

When the valet fetches his black Ferrari, she thinks they are going to his place. The car accelerates in an explosion of speed, growling lustily. Leo changes gears and turns up the music, finding sensorial pleasure from manipulating the machine to his liking. When it is finally just so, he thrums his fingers on the wheel and cracks a smile in her direction. To their right is the glistening black sea, breaking up the moonlight into tiny gold shards. A hillside park appears to their left. He stops the car near the entrance that says GROTTE DU LAZARET, jumps out of the driver's seat, and motions for her to follow him to a locked gate.

"This place is closed," she whispers. "We should go back."

"Where's the fun in that? Besides, I have a key given to me by a friend who works here."

He opens the gate and guides her inside, his hand resting on the small of her back. They walk for a while on a garden path leading up the hill. The outdoor sculptures of animals seem to be gazing eerily at them.

"Where are you taking me?" she asks.

"You said the world is ending. I'm going to show you that it's not." They stop at an entrance to a cave, also gated and bolted. "We're here."

Another key opens the lock, and mercifully the first thing he does is turn on the lights. He motions for her to follow him to the mouth of the cave, which is flooded by orange light.

In front of the stalactite walls, there are some pointy rocks that hardly look like anything—but video projections re-create how they were once used to kill and skin reindeer and woolly rhinoceros. Holograms of the earliest bipeds rise in Africa seven million years ago and follow the animals into Europe. The sun burns orange over the Alps, its slopes covered in a fine down of savannah grass. The grotto is right on the Mediterranean and fills with water as sea level rises. The Ice Age comes and the water goes down, exposing the grotto to the humanoids that shelter in it between hunting for elephants, hippos, and giant deer. In a snow-storm, the lions, the leopards, the spotted hyenas, the wolves, and the bears eye their human prey—but humans are dangerous; they have fire. The snow recedes again and the conifers are replaced by oak and palm. The water level rises and deposits another layer of sediment inside the grotto. The stalactite drips down over the bones of animals and humans, entombing them for more than a hundred thousand years. And on and on.

"So you see," Leo says, turning her to face him. "The South of France has been as hot as Africa and as cold as the Arctic over mil-lions of years. Yes, all those Paleolithic humans have gone extinct, and so have the animals. But the Earth is separate from what lives on it. No matter what happens to us, it will regenerate."

"So you *do* think we will all die."

She steps closer to him. He pulls her in by the waist and slides his hand underneath her T-shirt in one motion—there is neither tenderness nor urgency in that touch, only the feeling that they both know what they are doing and are watching from above. Since she offered him her painting, weren't they both in complete understanding?

"Death is what makes life interesting," he says, just before covering her mouth with his.

It's almost ten o'clock when she wakes up in her apartment. She makes coffee and sinks back in bed, replaying the scenes from last night in the cave. It was the kind of curiosity-based sex that she used to have in her early twenties when she knew, going in, that she wouldn't come. He didn't seem to care, either. But for what it was worth, they both performed the hell out of it as if the millennia of fossilized sediment weren't shredding their backs black and blue. It's always nice to get out there and have some unusual sex, she reasons.

She wonders if she should text Leo and realizes they haven't exchanged their contact information. Instead, she sees that Benoit has sent her messages late last night, asking how she's enjoying Nice. The glee with which she would have read his texts even yesterday is now gone, she notes with pleasure. Feeling lazy and empowered, she decides to wait to respond.

She puts on a newish bikini and a flattering dress and heads down to the cape, but Leo is not there. She cuts off swimming early and goes down to the flower market instead. The mimosas are in season, and people are walking around with bouquets of sunny yellow blossoms tucked under their arms or sticking out of their bags. Next to the flower vendor, she sees the Senegalese wood sculptor on his knees, picking up his elephants and rhinos. Behind him, three policemen are standing and gesturing bureaucratically at the wood shavings tossing around in the wind. The sculptor doesn't protest or show any emotion as he gathers his animals inside a garbage bag. No one seems to remark on his departure, not the flower vendors nor the fishmonger next to whom the sculptor had positioned himself for weeks.

She's no different—is she? She thinks that even her sadness is cheap; believes that she should at least be decent enough to not buy her way out with casual sympathy. Still, all the recent exertion to control her emotions—about the sculptor, Benoit, the world in general—makes her feel like a scuba diver taking measured breaths out of an oxygen tank. She can almost hear her own heart beating hard against the pressure.

When she sees the black Ferrari in front of her apartment, she feels more buoyed than she would have guessed.

"What are you doing here?" she asks Leo, peering into the front seat.

"I thought I would say hi. And see your other works, if that's possible," he says, only just lifting his eyes from his phone—as if *she* has come to *his* house to bother him.

But once inside the apartment, he turns out to be an appreciative visitor. She hasn't brought many tools and oil paints take too long to dry, so she's been mostly doing watercolors. She has felt increasingly drawn to the idea of synesthesia, so on her canvas the sound of waves dragging on pebbles and the peal of church bells have become aquamarine concentric circles enclosed in a yellow pyramid, and the resinous aroma of mimosas is a squiggle of green-earth paint.

"And what does emerald green signify?" Leo asks, pausing in front of a large piece washed almost completely in the pigment and nothing else. There's something in it that makes it impossible to look at anything else. It seems to exude its own radiance, even in the waning light.

"That one is a secret," she says.

Leo smiles and recrosses his arms. "You're a mysterious

woman. That's fine, keep your secrets. But even if you don't tell me what it means, I'd like to buy it."

"Are you sure?"

"I want to buy this piece, and all the other pieces too if you are okay with it."

"I haven't named my price," she says. "It could be very high."

"Try me."

She digs in her memory for how her pieces sold at the last exhibit in Greenpoint. She decides on a number halfway between those prices and what she thinks he would pay for a first-class ticket to one of his side offices. He nods, and nods again when she names the prices of the smaller pieces, and calls his assistant, Myrta, who will arrange the wire transfer to her account.

"It's probably not what I should say, but I'm still not really sure why you're so interested in my work," she says as she carries the tubes of rolled paintings out to his car.

"The short answer is that they inspire me. But it's more complicated than that." He dumps the tubes inside the trunk and slams it shut. "If you really want to know, come with me for a drive."

Theirs is the only car on the road to Sophia-Antipolis at six o'clock on a Friday evening. Sophia, meaning wisdom, and Antipolis, the Greek name for Antibes, explains Leo. And so, the research park is now the center of technology, bioengineering, and entrepreneurship in France, attracting companies and talent from all over Europe. As they park the car, bats fly by, swirling like a windmill over the darkening trees. They walk along a concrete footpath, and it dawns on her that his office isn't just a building, but a whole campus of white edifices tastefully arranged throughout the scrubby woodland.

In the last light of the dusk, she sees something cross their path and disappear into the bushes. She grabs a hold of his arm. "Did you see that, Leo? It looked like a fox."

"Yes, they live in these woods," he says, unfazed.

She shudders. It's colder here in the woods than in Nice. "But it wasn't red-brown, like the one I saw in the Alps."

"Because it's dark right now. Here we are—our lab."

He hovers his knuckle-size silver ring on the security scanner next to a glass door. It slides open as the interior lights up, revealing mid-century leather sofas and recliners positioned around a fireplace. Leo explains the rooms they pass through, each one sleekly furnished with monitors, wooden blinds, and Italian cantilever chairs. She doesn't see anything even approaching a lab until they reach a concrete door, where he hovers his ring over a scanner again. It opens soundlessly and they walk into a white room surrounded by glass. At first she thinks the walls give out to the forest outside. Then she realizes they are enclosures.

Leo smiles. "Sorry. I didn't want to give it away when you saw the fox."

In each enclosure, she sees an animal—a fox, a boar, a cat, a rat, an owl—sleeping or stirring restlessly. They are all the same color, which she has never seen before. It has never been seen before by *anyone*. If she had to describe it, she would say it falls halfway between green and pink. The beauty of the animals is terrifying.

"We've invented a completely new color in the history of humanity," Leo says, his voice gleeful.

"This is a color that can only be achieved by bioengineering on a living being. Just like something organic can't be metallic, and acid green doesn't exist in nature . . ."

"So why didn't you use insects, like carmine for red?"

The fox in its enclosure presses its snout against the glass, looking at her with irises of phantomic brilliance.

"We tried a lot of insects. And plants too, as a matter of fact. But none of those worked." He shakes his head, exasperated. "It appears that the pigment can only be created on beings with souls.

That's the element required for this chemical reaction—that, and warm blood. Like copper and arsenic."

The owl has begun to flap around in its enclosure, and she feels sick to her stomach. "So what does this have to do with changing human consciousness?" she asks.

Leo is no longer smiling his concealing smile. His impatience shows. He has gambled years of his life to this sui generis endeavor; and the moment he shows anyone some animals in cages—a very nice plate-glass enclosure, by the way—they accuse him of barbarity. But what he is doing is far more humane than creating a single hamburger patty. This painter, with her penchant for deadly pigment, *seemed* to understand the trade-off between pain and beauty—that everything sublime has a price. He had found that sexy, along with her gameness.

Now he thinks she sounds just like an American.

"This *is* consciousness itself," he says. "Don't you understand?"

When Benoit realizes he is going to be in New York for his cousin's wedding, he sends her a message. It makes him rather anxious—they haven't seen or talked to each other in ages—but she readily replies. Would she like to see him for drinks one day? he asks. Yes, she would.

Now he is wandering around SoHo, in the rough vicinity of the hotel bar that she has suggested. There is something strange here, he thinks as he loses himself in the cobblestone streets slick with a tropical downpour, amid the steamy smell of perfume and garbage. The hours seem to pass at a different pace, and time is as elastic as a mound of dough. He feels as though he has just woken up in his hotel room, but it's already late afternoon. This has happened every day of the trip, although he hasn't overwhelmed his schedule and has stuck to his usual habits.

After a while, he stops thinking it's the jet lag, this state of perpetual disorientation. To stave off the malaise, he had earlier made tea in his room and sipped it slowly, thinking of home. Despite the heat, he decides to order another one in a cafe buzzing with scents. Pu-erh-fect Purity is said to detoxify the brain of heavy metals. The young woman at the counter also suggests Peppermint Choco Chai-ngri-la, to help balance the chakras while celebrating Christmas.

As he holds a paper cup of scalding tea in his hand, sweat beading on his back from the humidity, Benoit realizes something. When the Americans shifted the months of the year—eliminated February and March, halved October and November and doubled June, July, and August—in accordance with climate savings, they actually changed the way time runs. Here-and-now in New York is not the same here-and-now in France, separated by a nominal six-hour difference, but a totally unsynchronized place-and-time, dilated like a clock in outer space.

Twenty years ago, when Benoit visited London, he'd heard the birds start singing at exactly 4 a.m. In France, however, they start singing at precisely 5 a.m. Of course this is because, despite the two time zones with a difference of one hour, the birds of England and of France recognized the single right moment to start singing. They were simply side by side on the same merry-go-round.

Quantum theory suggests that now, birds and people of New York are living on an entirely different merry-go-round. But not just them—*all* these places of the world have fractured off into separate realities. This is only observable to an outsider, and no one else streaming through the fetid streets of New York appears disoriented or oppressed as he is.

He means to ask her about this. Perhaps this is why he is here, thinking of her—someone from his past—as though she is in his present. It finally dawns on him that he feels sorry for the way he treated her back then. He knew how she loved him, or could love him, and pretended he couldn't see it. As if it weren't perfectly visible—impossible to hide, like love always is.

A scorching twilight descends outside the cafe. He stops walking when his eyes get caught up by a strange, ineffably beautiful color filling a shop window like a fog. It's displayed in the next boutique as well. Clothes, furniture, shoes, rugs, phones, and smart homes, all radiant in this ecstatic hue. In a chic ladies' boutique, there is even a fur coat in the new color, although it would be far too warm to wear almost anywhere in the world.

Benoit shivers as if he's having a fever and wipes the sweat from his brow. He reads the block letters stenciled on the window:

MEET KNIP, COLOR OF THE YEAR 202_. IT'S BLITHE, SEDUCTIVE,
YET INNOCENT, SYMBOLIZING HAPPINESS AND TRUST IN THE
FUTURE. THE COLOR OF SOULFULNESS, IT HAS BEEN CHOSEN FOR
ITS ABILITY TO GIVE PEOPLE HOPE.

It's now almost five o'clock, their appointed meeting time. He imagines her waiting for him at the hotel bar, but somehow finds it difficult to turn away from the shop window. He has the most irrational desire to bury himself in that coat. It would fill him with something he has lost forever—and he has lost much over the years. If he could just keep looking at it, he might recover the man he used to be, the man he once wished to become. The few people he had truly loved without cynicism. Frosted fall mornings that had yawned slowly over the mountains, glaciers on Mont Blanc, snow on Christmas Eve.

But now he takes a deep breath and wrenches himself away. He runs and doesn't stop until he's standing by the gleaming glass windows of the hotel. It's six minutes past the hour. He peers inside and recognizes no one.

For a split second, he thinks that she has stood him up, and relief floods him. He barely even remembers what she looked like that winter; *d'ailleurs*, she is now at that age when some women really change—not to say there isn't a certain appeal to middle-aged women, but they are like dried flowers, lovely in certain lighting and sad in others.

Then a woman at a corner table raises her head, and their eyes meet. He smiles and waves awkwardly. She's older, but still beautiful, thank god. She looks at him without smiling, judging him in return. He's forced to consider himself through her eyes, his grizzled skin and hair mottled with gray. She rises from her chair and disappears from view. So he has disappointed her more than she has disappointed him. What did he want, anyway? Validation of his lame ego, perpetually lurking inside like some nocturnal monster? He is so tired of the futility and vanity of it all.

His feet fall heavily on the pavement as holiday shoppers flood the streets. He has gone halfway down the block when someone touches his arm. She doesn't say anything as they hold each other in a brief hug, brushing their cheeks. She looks at him, he looks at her, and honesty passes between them and recedes like low tide. Some store is playing that one inescapable Christmas song by a pop diva, and the automatic humming of perfect strangers is hundreds of balloons rising together to the sky. Empty coffee cups knock between people's feet that step gingerly over the noxious body of a dead rat. Around them, everything glimmers in that mysterious color of the new world.

A WOMAN'S LIFE,
IN 10 SCENES

1.

The first time Celeste understood the word "black" was when she saw him idling by his horse outside the racetrack.

People thought that out in the desert, everything must be pitch-black at night. But the sky she knew was a lucid dark blue, all lit up by the moon which, not having any other competition, was impudently bright. Even on cloudy nights Celeste and her friends walked around in the range illumined only by the flickering ember of their cigarettes. Like little red stars fallen to Earth for bad behavior. There was nothing black in the desert.

But he—where to begin? There was his horse, long-legged and skinny like all the Reservation horses but armored in a lustrous black coat and daubs of red paint. It stepped in place and tossed its mane back in a gesture of pride and anger. Other riders about to race would have been terrified by such a restless horse. But he kept calmly stretching, keeping one leg straight while side-lunging deep into the other. His hair was braided into two long ropes down to his elbows, and their soft, black ends grazed the blooming dust every time he lunged lower and brought his crotch almost to the ground. Only when the horse unfurled its lips and whinnied in a desperate cry for attention did he turn around. He laid a hand on the horse's face without a word, and it soon stopped pacing and whipping its head. It was not quite the usual love and devotion with which horse people communicate with their beasts, nor was it discipline. It was something else altogether that quieted the bronco down. When Celeste's girl cousin told her the man was one Tommy Wolf, a Yakama, she instantly saw a black wolf that does things for its own reasons that not even other wolves can identify.

2.

Celeste was not just an average girl, to tell the truth. Her parents and most of her teachers expected her to graduate and even go to college. She tried harder at school than she let on and got more As than her friends knew. Celeste was a girl who once wrote an essay about a book on Frida Kahlo even though she'd never read that particular book. But there were so few books in Warm Springs anyway that it could hardly be her fault, and if there had been that book anywhere near her, she would have read it. In fact, she wanted so much to have read the book that she felt as though she'd done it, which was how she got to writing that essay. When it was finished, however, she felt a little ill.

Celeste played it cool in front of her friends, who had known her from the age of diapers only as Cece. Cece, a girl who could split a bottle of gin in an hour and climb quietly through her bedroom window like a coyote. Everyone else—people who didn't know her very well, sarcastic neighbors, and boys secretly in love with her—called her Miss Warm Springs, although she wasn't even eligible to compete at age seventeen. Her mother had held the title once and it was assumed that she'd carry on the family honor. That was one of the reasons people said the Thompsons were a bit uppity. Another thing was that her father was thought to be wealthier than he deserved to be. For these reasons the adult members of the Thompson family tried to keep some distance between them and their neighbors. But Celeste made friends with whomever she wanted, being the smartest and the prettiest girl in the Reservation by anyone's estimation. She was not the kind of girl who got easily scared, even by wolves.

3.

After the race, the sun burned down hard and fast; you could nearly hear the sizzle as it touched the horizon. Celeste was waiting for her cousin outside the gates when Tommy Wolf came out, leading his horse with one hand and fist-bumping his holder with the other. He had won, just as she'd predicted, and he was in that cloudlike euphoria of winners where only other winners were granted admission. He didn't turn around to look at her, even though she was glowing rose and violet in the twilight. Just as she'd predicted.

4.

Warm Springs Reservation was formed by the Treaty of 1855 whereby the Wascos and the Warm Springs tribes (the Tenino and the Tygh) ceded ten million acres of land to the U.S. government in exchange for $150,000 and 578,000 acres for their exclusive use. Historical economists say that the Indians were compensated for the 9,422,000 acres at roughly $0.99 per acre in today's terms. They also point out that it is a common misconception that the Reservation was in Indian ancestral lands. "The place you have mentioned, I have not seen. There is no Indians or whites there yet, and that is the reason I say I know nothing about that country," the Wasco chief Joseph Mark said at the time, plaintive in a very understated way. The Wascos were once a prosperous river tribe that controlled the commerce in mid-Columbia River. When they arrived at their new Reservation, they found canyons and plateaus clad in red soil and rabbitbrush that was good for naught except bands of wild horses.

This was, of course, completely intentional on the part of the U.S. government. They had worked hard to give the Indians the least desirable land in Oregon, and for a while it looked like they'd succeeded. But what the whites completely missed at the time was that there were two springs in Warm Springs: an ice-cold and clear one that fed into the Deschutes River, itself a tributary of the Columbia; and another diabolically hot and healing one where you could bathe comfortably even in the middle of a high-desert blizzard. The use of the cold spring was shared equally by the community in this arid land. But the right to the hot spring, which was primarily valued for its sacred uses, was divided between Chief Joseph Mark and the respective chiefs of Warm Springs and the Paiutes. The lines of the other chiefs soon died out, so the oldest heir of Chief Mark became the sole controller of the hot spring by 1905. This did not bother the rest of the Indians as long as the Chief's heir continued to allow them to bathe. Then in 1965, Flora Mark—the last of Chief Mark's bloodline—married one Randy Thompson, thereby handing over the hot-spring rights to another name.

By this time, the electric company from Portland had dammed

up the Deschutes at a few key places so that the salmon could no longer swim up to their nesting spots. To make their chances even worse, the company eventually installed hydroelectric turbines to generate power that was ferried more than a hundred miles to the city. That must have meant money, although most people never saw any—all they knew was that the salmon were gone. It was widely and wisely speculated that the whites would come after the hot spring next, and its sole owner would die a very wealthy man. This was how Flora and Randy's son, Randy Jr., became the object of collective enmity, other than the fact that he married the prettiest woman in the Reservation, the former Miss Warm Springs.

5.

Celeste's cousin had a friend who lived in Walla Walla, who invited them to the relay after-party at someone's house. It was in the middle of a hay field just outside town, the kind of place you recognize by the crunching of the gravel driveway and dogs barking in the distance. There was a fire going in an oil drum in front of the house. Around its warmth, guys were clutching beer cans and girls were rocking their bodies to the boom box half-hidden by the long blond grass. This was familiar to Celeste, but she nonetheless felt intimidated that most of them looked older than her. Her cousin, who possessed a worldliness belying her eighteen years, strode up to the ice box and fished out two cans of hard lemonade. Almost as soon as they took their first sip, some huge guy swooped in to cage her cousin in a full-body hug. Beneath the torn-off sleeves of his T-shirt, his biceps were thicker than Celeste's thighs; despite the darkness, a pair of wraparound sunglasses was perched atop his head. The cousin laughed; this was the friend whose party this was. *I have to catch up with him*, the cousin whispered. *Go see if Tommy Wolf is here,* she added with a wink.

She had already scanned the group around the fire immediately upon arrival and not found him. There were some more guys sitting on plastic lawn chairs on the porch; they gawked at her as she went through the open front door. It was hot inside with the smell of beer and sweat, with that particular excitement of bodies—the way, especially when one is young, the mass of other people's bodies in a room can offer a world of possibilities. Celeste took a gulp of her lemonade, told herself to play it cool. There were riders here, the guys still wearing bandanas and war paint and long socks with cut-off denim shorts. Girls were fanned around them, giggling and bumping into the guys and acting drunker than they were. A short girl with tiny denim cutoffs and waist-length hair was burrowing her head into a rider's chest. He had black braids and the hand that he raised to lightly embrace the short girl was uncharacteristically calm. It was the same gesture he'd used to quiet his horse. Celeste's heart dropped to the floor so hard that she was half afraid other people could hear it. She was certain they could at least sense it, because he looked at her just then. He unwound himself from the short girl. Turned and looked at her again, a little bit longer this time. Celeste gulped down her hard lemonade. Third time their eyes met, he disentangled himself from the circle and walked toward her. There was a bit of idleness in his approach too, just as when he was stretching before the race. Not fast, but focused.

6.

It was rather unjust that Eva Thompson had a reputation for being uppity because she worked at five different jobs only to get by.

She sewed; made beaded bags and ornaments for the sole souvenir shop in town; baked huckleberry pies for the stall by Highway 26 that Randy ran; worked behind the cash register at the local grocery; and did accounting for the antidrug association. All she'd ever done was save enough for Celeste's first year of college, if she went to a state school. After that, it was hoped that Celeste would cobble together student loans, scholarships, and part-time jobs to put herself through the remaining three years. But people still said Eva "gave airs." Such as, the decades-old thirty-two-volume *Encyclopedia Britannica* that she'd bought from a traveling salesman as a young bride (foreseeing that her future children will need better books to read than what the Reservation school could offer). Such as, her insistence on painting her nails and blow-drying her hair. And the dreamy way her eyes drifted to the horizon while pretending—badly—to listen to the rez gossip. It was not exactly said, but universally felt, that Eva acted like someone destined for a better place. Randy told their daughter, Celeste, that Eva "had an artistic side."

Because of this, no one was surprised when Eva declared her wish to learn to code. Ostensibly she wanted to make a little extra for Celeste's college fund, but her family recognized that it was equally the latest burst in her lifelong quest for self-determination. She drove out to Madras with Randy and brought home a new computer like a prized puppy. Every night, she sat in front of the computer (installed on one end of their long dining table) and an open copy of *Coding for Dummies*, occasionally checking her work against a YouTube tutorial. Sometimes Randy gave her shoulder rubs and Celeste offered her tea, whispering words of encouragement, as though she were running a marathon and they were her coaches.

Mostly, however, they left her alone. Randy, for one, was busy dealing with a company called Vesuvius that wanted to talk to him about the hot spring. He had the invigorating feeling that he'd been waiting—preparing—for this moment all his life. He thought that he would intone a dignified "No," the whites would beg and implore, he would never waver, and then they would be sent packing with their tails between their legs. But then things got complicated.

"We understand your concerns, Mr. Thompson. Especially in light of the historic exploitation of your people's land and resources. Please understand that you have our fullest sympathies and support," said the director of business development. He was neither young nor old, with black-rimmed glasses, carefully side-parted hair, and a red-and-black plaid shirt that was supposed to be rustic but couldn't be bought anywhere outside a city.

"Not sure if you drove up the Deschutes, Mr. Chapman," Randy said, and the man squealed, *just call me Dax!*

"Okay, Dax. Not sure if you saw that dam there, but after that was built we stopped having any salmon runs for fifty years. With the fish ladder they've returned, but they still have a hard time coming all the way up. Our people have been eating salmon for ten thousand years, you understand? We gave that up for what— power in Portland? Anytime we open ourselves up to you folks this is what ends up happening. It's the Indians that get short shrift."

Dax nodded with the solemn air of someone whose grandparents have just passed away. "I hear you, Mr. Thompson," he said breathily. "Please know I have the deepest respect for your way of life . . . I absolutely love nature . . . I go hiking, mountain biking, and rock climbing every chance I get . . ." Dax cleared his throat.

"But you should know, Vesuvius *is* what's good for the planet. We've developed the world's first-ever scalable technology for a geothermal energy plant that is—we believe—*the* key to mitigating climate change. Clean, pure energy with zero carbon emissions. A hundred percent renewable. And with profit share in perpetuity.

This is . . . Something bigger than Indigenous or white or POC. This is about the future of Earth!"

After Dax left, Randy sat still on the aging sectional permanently indented in the shape of the family's bodies. Here was Eva's bottom on the chaise longue. Here was Celeste's nook, where she liked to read. Suddenly its dilapidated state came into his eyes like never before; that sectional had belonged to his parents, for god's sakes. The money that they were offering would be enough to buy thousands of new sofas. They'd never have to work again, Randy and Eva and Celeste, and even Celeste's children. In perpetuity.

"Celeste!" Randy called out, and his daughter came out from her room, rubbing her eyes with an air of exhaustion. "I'm going to the river to count the fish. Wanna come?"

7.

Counting the salmon and steelheads was something Celeste had done with her father ever since she was little. As usual, she stood by the riverbank and looked for the fish carcasses bobbing in the current. Their spawned and spent bodies were like logs of silver against the darkly flowing water. The pearly remains didn't make her sad, however. They had succeeded in coming home. Just one out of a thousand brothers and sisters made it back to the stream of their birth. Even when the water became too hot and too low year after year, even over dams, turbines, and runoffs, they insisted on fulfilling their life's purpose. They seemed to be saying, *Some day when the rivers dry up, we will dry up too—but not until then. Not yet.*

So the more dead fish were found, the happier she was. Only when she saw the eye of a half-rotten, fuchsia male Chinook did she look away in queasiness. Although she made a principled attempt to remain indifferent, the nausea was a constant reminder of that July day. The sun and dust and horses. Fire in the oil drum, his hands on her body. Their surprising warmth and gentleness. For a while he had texted her every day, then every other day, then once a week, until she lost all sense of the brief, dizzying happiness of wanting and being wanted. Still, she wasn't angry. She believed that he would drive down to see her from White Swan if he knew how she was struggling on his account. But he hadn't mentioned meeting up again and she was too proud to use her secret to force him.

Her father shouted, "Thirty-two!" He sounded revitalized and happy. When they got home, he told Eva what had happened with Vesuvius and that he was going to refuse their offer.

"You know I support you, honey," Eva said, turning away from the computer screen. Then with that dreamy look in her eyes, she added, "Did you know about this thing called Bitcoin?"

Randy hadn't heard of Bitcoin. So it fell on Eva to explain from the beginning that there were these things called "cryptocurrency" that were essentially "bits of long, unique code" that were so indecipherable as to be worth money—in fact, lots of money. There were coders, really smart people all over the world, who were "mining" these bits of code like gold coins from the digital ether. That's what they did for a living. And one of them, whom she met in an online forum for self-taught coding students, offered her a special deal on some of his Bitcoins because he valued her academic fervor, the way she was trying to rise above. At her age, no less.

This didn't make sense to Randy from the very beginning, but he asked hopefully, "What kind of a deal?"

"I bought a hundred Bitcoins for ten thousand dollars," Eva said with pride. Randy groaned, and Celeste crossed and uncrossed her arms.

"Eva, that's Celeste's college fund. Everything we've saved. Are you out of your mind?"

Eva laughed, and at that moment her face had the radiance of a former beauty queen. "Honey, you're upset because you just don't get it! The thing about Bitcoin is that the prices go up, up, and up. In six months, we'll have enough to pay for the four years' tuition. You'll see!"

"Okay, then what is it that you *bought*, exactly?" Randy connected his words together with difficulty. Eva went swiftly to the family safe and brought back ten sheets of paper. Celeste took the first sheet, which said this:

```
10 Bitcoins, National Australia Bank
<aspoijac92835709qpjsfdaonaklnsdl
ty18u23u50q9jajsmzljdf8qqiw05utojspoajosdf
japosjoziz==1390234h?/0293588!>
```

"Mom, I don't know," Celeste said.

"This is the code!" Eva shouted, a bit more impatiently this time.

"This doesn't look real, Eva." Randy dragged a hand down the length of his face. His cheek stretched like putty under his palm. "How did you pay for this? Maybe we can get the money back."

"I wired it . . . William said he's in Australia!" Eva insisted, but her face was turning white. The next morning at 9 a.m. sharp, Randy walked into the bank in Madras with the sheets of Bitcoin in his hand. The branch manager looked at them carefully and brought over his associates to his corner office. Soon they called the police, while Randy asked again and again, since they had the account number that had received the funds, couldn't they just simply reverse the transaction? They did trace the account, not to Australia but to Kerala, and it was completely withdrawn and closed already. The bank said that since Eva had willingly made the wire transfer, this was not a "fraud" but a "scam," an important distinction which unfortunately put her outside the protection of insurance. The police said that if this had happened domestically, it would be a different matter, but pursuing overseas criminals was nearly impossible.

8.

Celeste's mother stayed in bed for a week, moaning and twisting fistfuls of her sheets. But on the seventh day, she got up, took a shower, and started rolling the dough for the huckleberry pie. As a way of explanation to Celeste, she only said that it was like she was enchanted. She thought it was a lovely dream but when she woke up, it turned out to be a nightmare. Before long, her mother was back to her usual self; the only difference was that she no longer painted her nails.

Most people would have thought that in light of the Bitcoin incident, it made sense to take the offer from Vesuvius—at the very least for Celeste's sake. One door closed, but another remained open; Celeste could still become one of the few seniors to go to college that year. But Randy interpreted the signs completely the opposite way, that outsiders cannot be trusted and that vying for quick money always brings ruins. So he called Dax Chapman and refused the offer, certain that his neighbors and relatives would appreciate his moral fortitude. But when the word got out about Vesuvius, they were angry that Randy didn't sell the rights and share the profit with the rest of the tribal members. They said he was holding out for a higher bidder when he knew what good that money could do in the Reservation. Selfish, they said. Greedy.

Even Celeste's longtime friends acted cold around her in a way that suggested they had turned their backs on all the Thompsons. When the few remaining sympathizers invited her to their parties, it didn't go unnoticed that she'd stopped accepting drinks or cigarettes. She learned it's easier to stay home to avoid getting questioned. For the first time in her life, Celeste didn't have anyone to talk to, so much so that she caught and stopped herself a few times from talking to *it*. That's when she abandoned her principled resistance and texted Tommy, and he drove three hours from White Swan to see her.

Their first meeting since the relay day was at Reused Cafe. Celeste waited in one of the four Formica booths surrounded by racks of used clothes, baby stuff, DVDs, and gently broken furniture. Her eyes leaped up when he came in. The same long black braids. The same intentional walk. When she stood up to hug him, he caught her in his arms and tenderly lowered her down as if she would break otherwise. This made her so happy that her eyes became hot and red.

"Tommy," she said. That was all she could say in the moment.

"Yes, Celeste." He reached across the table and grabbed both of her hands. "I'm going to get a coffee. Can I get you anything?"

By the time Tommy came back with their drinks, Celeste was feeling hopeful. Maybe this was how it was supposed to be, getting taken care of. He watched her as she took a big sip of her latte.

"How are you feeling?" he asked. "Is everything . . . okay?"

"I don't know. I haven't been to the doctor," she said.

"Then . . . Are you sure?" Tommy bit his lips. "Maybe—"

"I took two tests," Celeste said. She'd also missed her periods and thrown up all her meals, but she didn't mention this. Tommy nodded and stared at his hands, interlaced together on the table.

"So, what do you want to do?" he asked. His coffee was turning cold untouched, she noticed. What did she want? Did she still have a chance at going to college? Or a chance at Miss Warm Springs? What would happen if she kept it? She would become fat and ugly, and Tommy would lose interest in her. Her parents and teachers would be so disappointed. Maybe this ignominy would put the final nail in the coffin of the Thompson family.

"I don't know. What do you think?" she said. Tommy sighed.

"I was going to tell you. I'm moving to Portland," he said, eyes downcast. "You know there's nothing in White Swan for me."

"What about your horses?" Celeste asked, and the way he squeezed his eyes shut made her regret her question immediately.

"I've sold them already," Tommy said quietly. "It's for the best. Sometimes in life . . ." He smiled in lieu of finishing his thought, but Celeste was glad he left it unsaid.

"So I'll be in Portland, searching for a job. It will be hard for me to look after you. Do you know what I mean?"

"Are you saying I should get it taken care of?" Celeste's voice quivered. Tommy took her hands again in his own.

"No, that's not what I meant. We still have some time to think about all this, right? Maybe I'll find something quickly in the city, and who knows . . ." Tommy's voice dissipated but she knew he was trying. Maybe he'd get a job soon and ask her to join him; maybe she would also look for work while waiting for the baby to be born. Now she realized she'd always called it "baby" in her head without meaning to. The next thing she thought of, with no rhyme or reason, was a coffee table, which she thought was important to have in a grown-up apartment. It made her feel a little proud that she had foreknowledge of these things, which Tommy would have no notion of now but would appreciate in time.

A month passed after the meeting in which Celeste patiently

waited for Tommy to send word. A call, a text, a smoke signal. There was nothing. As another month passed, Celeste found it increasingly hard to squeeze into her normal clothes. In a fit of anxiety, she confided to her girl cousin what had happened. She, in turn, reached out to her Walla Walla friend, who was close with all the horse-relay crowd in Yakama. Evidently, Tommy was working at a gas station in Portland, somewhere near Burnside—according to a friend who had visited. This was deemed a good enough lead that Celeste decided to skip school one Friday and drive up there.

He hadn't answered any of Celeste's messages in weeks. But when she texted that she was on her way to Portland, he wrote back immediately saying he will meet her. Around 8 p.m., when his shift ended. Right now he was busy at work. *Where?* She asked simply to avoid taking too much time away from him. No response. Celeste turned her eyes back to the open window where the plains rolled on in silence. The sagebrushes with their yellow-blooming fingertips were holding aloft a dense November fog. Hidden by the atmosphere, at a place due east from Celeste's car, the hot spring was gushing out into the open, its steaming waters once again forgotten. The sky was an undulating gray from horizon to horizon. There was nothing; there was everything.

She kept driving along Highway 26 until the high desert turned to a dry juniper forest, then a rain forest of Doug firs and cedars. The trees were replaced by country stores and then strip malls and eventually, an unbroken stretch of tattoo shops, super-markets, restaurants, theaters, hospitals, and especially gas stations. There seemed to be one every other block. In Warm Springs, there had been only one.

Where is your gas station? she texted him. No response. It was still only 5 p.m. At this point she hadn't realized what she'd have to endure, because she was wondering how she might spend the next few hours until their appointed time. It even occurred to her that she might be able to catch a movie or eat something, after first finding his gas station. No response. But when she got to Burnside, she found out that it was a very long street that ran through the entire city with seemingly thousands of gas stations around it. After stopping at ten gas stations and asking if a Tommy Wolf worked there, she parked the car and decided to walk.

The sun had set hours ago. No response. It astonished her to see how dark it was all around; the streaming lights made the rest of the world—the in-between places—look even darker. Every step she took drew her into the shadows, yet she felt powerless to do anything else. It got windier and colder. No response. Her vision blurred from the gale; she kept walking, wiping her eyes with the backs of her wrists. When she opened them wide again, she realized she was standing on a bridge. Even its rusty red steel looked sable on a moonless night. But that was not when she realized the true meaning of the word "black." It was when she stared down on the lapping river below that she saw clearly the blackest things in the world—the human heart and the watery depths in which the fish floated and the watery depths inside her where a human heart was beating in complete darkness. She felt a strong yearning to bob along in the current like a silver log. She would be embedded into the black, he would have to remember her then. The railing only came up to her waist, and she leaned out into the night. Just before she let herself go, her phone vibrated in her pocket. Not Tommy. Eva.

"Where are you, honey?" her mother's voice said. "I ran into your teacher at the market and she asked if you were sick today..."

Celeste didn't know where to begin. They would not understand.

"Mom . . . I don't know where I am," she said, and the terror and sadness she'd dammed up inside came to break her down all at once.

"Can you come get me?" she said through sobs, holding on to the rails. *I'm so sorry baby. Baby I'm sorry.*

"We're in the car now," Eva replied. There was a sound of her directing something to Randy, car doors opening and slamming. "We're coming to get you."

No matter where in the world.

9.

Frida Kahlo, born Magdalena Carmen Frieda Kahlo y Calderón (July 6, 1907, Coyoacán, Mexico–July 13, 1954, Coyoacán), was a Mexican painter best known for her searing self-portraits exploring themes of identity, the female body, sexuality, and death. Kahlo was also known for her stormy relationship with artist Diego Rivera (married in 1929, divorced in 1939, and remarried in 1940). Kahlo's father was a German of Hungarian descent, and her mother was a Mexican of Spanish and Native American descent. In 1925 Kahlo was hit by a bus, which so gravely injured her that she underwent more than thirty operations throughout the course of her life. During her recovery, Kahlo taught herself painting by drawing all over her full-body cast. Although she passionately loved Rivera, their marriage was not a happy one. They each carried on numerous affairs (he notably with her younger sister Cristina and she with both men and women). Kahlo appeared to have transcended the many betrayals of her philandering husband: "Perhaps it is expected that I should lament about how I have suffered living with a man like Diego. But I do not think that the banks of a river suffer because they let the river flow, nor does the earth suffer because of the rains, nor does the atom suffer for letting its energy escape. To my way of thinking, everything has its natural compensation." But inwardly she continued to struggle even while gaining international celebrity as an artist, a Socialist, and a bohemian. Further casting shadows were her miscarriages, after which she painted some of her most harrowing and powerful works. Kahlo did not sell many paintings in her lifetime; the most money she ever made from her work was through *The Two Fridas*, which was acquired by the Instituto Nacional de Bellas Artes in Mexico City for around $1,000. She had only one solo exhibition in Mexico in her lifetime, in 1953, just a year before her death at the age of 47.

—*from Encyclopedia Britannica*

Nothing is black—really nothing.

—THE DIARY OF FRIDA KAHLO

10.

The late July sun was igniting the skin of her thighs on the stands, the rippling backs of horses, the slightly sweet-smelling sand. Her eyes followed one rider only, ever since seeing him for the first time that morning. His starting mount was a shining chestnut mare with lightning bolts painted on her rump. In a single effortless movement, he leaped onto her bare back. No saddles or stirrups were allowed in an Indian horse relay. The announcer introduced each rider and his team, and when Tommy Wolf of Yakama Nation was called, Celeste cheered wildly. The other horses paced in place but not this chestnut mare—she was dead still and calm like her rider, the perfect first horse to start the relay.

The flag went up and all the horses bolted as one. Tommy's chestnut instantly stretched her impossibly long, strong legs at a frightening speed. *And in the lead, there is team Black Wolf representing Yakama Nation in the black and red, followed by Thunderbird from the Crow Tribe in Montana!* the announcer shouted. *But it is too early to tell. Just now coming to the end of the first mile . . . Here they are for the first exchange! The mugger has to literally get mugged by the horse arriving at full gallop and the holder has to give away the next horse without a hitch. This is where the winners and the losers are made!*

Tommy rounded the oval in the lead but as he jumped off the first horse and ran to the next one, Thunderbird went whooshing past and almost knocked him to the ground. Tommy's second horse was a piebald wearing a wolfish mask. It reared up almost vertically while Tommy clung to its back with his thighs. *Black Wolf having a hard time there as Thunderbird takes the lead and Sioux Strong sneaks in at second place!* Celeste clutched the edges of her seat and groaned. *And Mountain River in the lime green takes the opportunity to get in third place. Black Wolf of Yakama Nation is now starting the second exchange in fourth place! This unpredictability and danger is really the beauty of the Indian horse relay, right Bryan?*

Tommy finally got his piebald under control. His face was set in stone—but he was now whipping his horse like he hadn't done in the first lap. The piebald roared forward, overtaking Mountain River within seconds. Its legs were just as long as the chestnut's, but it had a broader chest and neck that propelled its weight onward like a torpedo. *The third and final mile coming up for these warriors. Thunderbird is in the lead for the last exchange. AND we have a collision! Sioux Strong ramming straight into his own next horse! The mugger and the holder are desperately trying to gain control of both horses. That's going to be hard to recover from! Now there's Black Wolf coming in for the last mile!*

Tommy barely slowed his piebald before leaping off and alighting onto the bare back of his last horse, the black bronco. With no urging or hesitation the bronco shot forward like a scream. *Wow, now* that *was a beautiful exchange! Black Wolf trying to make up for lost time and pulling into second place, just behind Thunderbird.*

In a way it looked like the black bronco was simply floating and the earth was moving beneath its feet. Celeste realized that the horse had been angry earlier because it was forbidden to run. Now it was flooded with happiness in every part of its horse brain and horse body. *Thunderbird and Black Wolf are now running exactly side by side. Thunderbird on the inside, Black Wolf on the outside. If you're watching this on the big screen it's hard to appreciate just. How. Fast. These. Horses are going! They are galloping full throttle, ladies and gentlemen. This is the last half mile and both teams are tired. It is now a battle of wills.*

Celeste screamed, then the whole crowd jumped to their feet and drowned her out.

HOLY COW, Bryan! Black Wolf pulling ahead in the last quarter mile! Where is this energy coming from?!

Thunderbird looked to the side and ferociously whipped his horse, but it was utterly spent by now. Tommy and his black horse looked unfazed, like they would soar to the ends of the earth.

Aaaaand Black Wolf representing Yakama Nation in Washington wins the race! Ladies and gentlemen, our new champion!

Celeste was shaking from seeing something that wasn't a sport or even art—this was something that simply *was*. Which was enough to make everything else disappear. How could she ever go back to not knowing this?

Tommy took his whip in his teeth and raised both arms above his head as the horse still galloped on. His long black braids streamed behind him. Beyond the finish line, beyond the racetrack, they kept flying to stop the world from moving.

KWAZULU-NATAL

When it's blimmin hot like this in August, during dry season when it's supposed to be cool but the sun makes the yellow earth go poppoppop as it cracks, I always think on my elephant. It was just this kinda day when I was eight when my da brought him home in the back of his bakkie. Da pulled over and went round and lowered the ramp for him to walk down on, but he didn't want to move. Da said Jezus it's bladdy hot I'm going inside, and just like so he disappeared. And I got so worried leaving the elephant by itself in the bakkie with the sun beating down on his back, but I ran inside chop-chop and got a bucket of water and an umbrella. I climbed on the bakkie and the elephant still didn't move or make any sound like it was dead. It was a few inches shorter than me and its trunk was the size of my own arm, a baby probably born in the last wet season.

Come on, little oke, I said, setting down the bucket in the front of him, be a good elephant and drink some. He stood still there unmoving even tho there were goggas buzzing round al-lasudden attacking him. He didn't even try to flap his ears to get the bladsuckers off and cool down a little. I sat down next to him shooing at the flies with one hand and holding the um-brella over us with the other until the sun went down and the stars came up.

Da came back out and said, What are you doing there and not making the dinner hey, and I said, If I go away he'll die. It was by now freezing cold and I was throwing my body over the elephant like a blanket. He started coming to, only cos he was shivering and not cos he was done being bladdy stubborn. We were both shiv-ering like so and Da said You better check yourself for the ticks when you come in, I'm ganna beat you raw if you don't. But then Da threw me a blanket and a vetkoek before going back inside. I put the blanket on Rocky—Stallone was the tops back in day, bru—and I was scarfing down the vetkoek with both my hands when I noticed a plunking sound and realized it was Rocky dip-ping his trunk in the water bucket.

Ever since that moment me and Rocky did everything to-gether. When I ate he ate, when I sat he sat right down on his arse, and I slept cuddling him in the kraal Da made next to our house so I could feed him his bottle in the middle of the nights. While he sucked on it he liked to snuggle his trunk in my armpit, it made him feel comfy like being with his ma. We both didn't have mas, Rocky and me. My ma I have no recollection of cos she ran away when I was a baby, but I know she was Zulu. My da was Afrikaner and he was a hard, hard man, hey. He drank also, but he was just a bladdy mean kind of oke. Even his chommies, other rangers at the park, were kinda spooked by Da sumtimes. Nobody bothered him for having a bruin ou for a son hidden away on the fringe of the park, far from the other rangers' houses, cos they all knew Da would kill them and their families if they so much as laughed the wrong kinda laugh at him.

That's why they put Da in charge when they had any kinda culling mission to do. That year it was the elephants and the park decided it had more than five thousand elephants in excess of. They were ganna kill two hundred a year until they got the numbers down to where they wanted. So that's how Da and these other okes tracked down one herd of thirty or so which was picked out of some bad luck. They snuck up in their Jeep, and Da aimed and shot the matriarch right between the eyes like he's supposed to, killed it instantly. Then all the other elephants were confused and standing still, so aggrieved they were at losing her, and Da and his mates went and shot them all down cos if you leave even one alive it's ganna be trouble later on.

Afterwards they were having a pack of smokes back at the Jeep all quiet cos they were in that special zone of having done something they don't like only out of a sense of a force bigger than them, hey. The smokes was that camp of unsaid acknowledging between them that they did something terrible but necessary that okes sumtimes had to do to keep the world turning like so. Once they were finished Da said I'm ganna go make sure they're all dead and got off with his rifle. He walked round where the thirsty yellow earth was drinking in their red blad and their bodies were already giving off the sick, shitty smell of death. There was nothing moving except huge black goggas tearing into the wounds, laying their eggs already. But then there was summin slightly off, Da saw the leg of a large cow shudder a bit. Hidden beneath there was Rocky, hiding his face in his ma's haunches, his ears stuck flat to the sides of his head.

Da put his rifle up, aimed at Rocky, and then—I reckon this is why you never know what kinda oke anyone is, hey—he put it away. It was against the rules to keep the babies alive, it made them grow up to be bad elephants. All Da needed to do was leave him there and let the nature take its courses, the lions or the hyenas would have eaten Rocky like a sarmie before dinnertime. I don't know why to this day, but Da brought the Jeep round and got his chommies to help him put Rocky in the back.

After bringing Rocky home Da wanted nothing to do with helping me raise him. It was all up to me to feed Rocky every few hours and play with him. Da taught me how to read and write but I never gone to school—he didn't want to send his son to a school for blacks. So I never had any mates or anyone to talk to except Da until Rocky came. I was his ma, boet, and chommie all at once, hey. After I started to sleep inside the house again, this was how our day was. I woke up before sunrise cos Rocky came round and stuck his trunk inside my window, sniffing for me. We went for a walk in the bush staying pretty close to the house like Da warned but getting a bit farther out each day, Rocky being curious to see what was out there. He liked to chase warthogs or impalas that come in our path but when he see a cheetah he hide behind me. Rocky knew exactly what to eat even without his ma, he went right up to the umbrella acacia and munched on the finger-long thorns like cendy. Back home we played football or tug-o-war with a stick he found. At night I had to tuck him in, putting my hand in the soft snug place behind his ear or he couldn't fall asleep. In the morning it would begin all over again. I can still feel our walks, my hand on the spiky skin of his back and the calm flapping of his ears shaped like Africa, the rustling of the gruss, the sweet smell of acacias weaving through the icy dawn air. That was the best ever, my bru—I can't ever forget. It was like we were one and each knew what the other was feeling. He knew right away what was wrong with me, even sumtimes when I was sick he got all worried for me.

One night—this was about five years after Rocky came—Da drank too much and got all mean. He said, You care about that elephant more than you care about your own Da, izit? I shook my head. You bladdy liar, he kept going. I should've killed him when I was supposed to. This made me mad and I shouted, You have to kill me first before you kill Rocky. Da looked at me like he couldn't believe what he was hearing and said quietly, Come here. I didn't budge and then next thing you know his fists were flying at me left and right, right and left, sort of like Sly Stallone blikseming his punching bag. I stumbled backward out of his reach, which usually stopped him, but this time he kept coming after me shouting, You bladdy liar, I'll kill you and then I'll kill that fucken elephant too.

Summin came over me and I just burst out the door and ran into the pitch-black bush, hey. I ran and ran, not realizing how my skin was being torn to shreds from the acacia thorns. When I finally stopped I saw in the moonlight that blad was soaking through my clothes like I stood outside in a downpour. It was silent all round me. I realized allasudden that Da didn't follow me—and that instead he mighta gone to the kraal to shoot Rocky. I was about to turn and run back, make my feet fly, when I saw these bright glowing eyes in the bushes to my right.

If you're ever unlucky enough to run straight into a pride of lionesses, my bru, just know that the worst thing you can do is run cos then they know for sure you don't have a rifle or anything. So I backed slowly away from them, and they just as slowly came closer to me. There was a noise like the ground being torn up, trees breaking, and I thought, Here they come. But it was coming from behind me. The earth was boombooming now. And through the trees behind Rocky came bursting out, raising his trunk high over his head and trumpeting as loud as he never done before. That

drew the lionesses out into the moonlight and there were three
of them—I knew them by reputation. Three sisters with three
little cubs each that needed to be fed, and a baby elephant that
Rocky was then would have made a lekker meal. But Rocky—you
should've seen him. He charged at them, stopping and making
sure I was still behind him, rumbling, then charging again, until they gave up and slinked back into the bush. Rocky just knew
I was in danger, and he came to save me, hey, he and I had that
kinda bond where we were in each other's skins.

Not long after, Da said Rocky had to join one of the herds in the
park. I knew it was the right thing to do, for him to be with his own
kind eventually. Da told me where to lead him on our morning
walks so he could get to know these elephants. When we saw them
drinking and bathing in the watering hole, I said to Rocky, Go on,
little oke, go make new chommies. If you get scared then you come
right back to me. Rocky wrapped his trunk around my shoulder
like he wanted to stay and I peeled him off me, saying, Come on.
He was curious too, so he shuffled forward swaying his trunk all
shy. When he poked out of the bush all the elephants looked at the
matriarch, trying to see what she do to Rocky. She took one look
at him, turned round and went back to grazing. Seeing that, the
other elephants went right back to doing what they were, ignoring
Rocky. He walked a little closer to a calf about his size, trying to
say howzit. But that calf's ma rushed forward and whacked Rocky
with her trunk so he slipped and fell in the mud. She rumbled and
kept pushing him this way and that until he got his legs beneath
him and ran back toward me, shaking. We kept trying for months
to get Rocky accepted by a herd but nobody wanted him. If he kept

wandering round the bush by himself the other elephants mighta killed him.

Rocky kept growing and needing hundreds of kilos of food every day. Da told me he got his boss to arrange for a transfer to a private reserve nearby—it was a place that didn't cull so he be safe. I slept with Rocky that night in the kraal, putting my arm in the soft place behind his ear as we lay. The next day, I got on the back of the special bakkie for elephant transport. He didn't budge until I said, Come on, Rocky, don't be scared, I'm going with you. We drove for maybe an hour and a half south in the soft, misty rain that glows off the trees like halos. When we got off I read a sign that said *U__ Sanctuary* and then another one saying *Elephant Interaction Tour*. I said to Da, I hope Rocky gets to make lots of chommies here, and he said, There's only one other elephant here. I said, In the whole place there's just one other elephant? It turned out that the sanctuary was separate from the reserve, it was only a closed-off place for animals that got raised by humans and couldn't go back to being wild animals. Cos Rocky was so tame they were ganna put him in an exhibit from morning until late afternoon, seven days a week. Fully knowing this and how much Rocky was ganna hate it, I led him to his new kraal and fed him his favorite dried corn one last time. Rocky my boet, this isn't the end, I whispered to him, I'm ganna come get you as soon as I can. Rocky snorted up the kernels through his trunk and waved his ears at me like he thought I was just going to the toilet and was coming right back.

After that, I couldn't stand being with my da anymore so I begged him to let me get a job. I was not quite fifteen when I started working at a coal mine near Richards Bay. For the next fifteen years my life was complete darkness, I got just one day off each month. Buried beneath the rocks I kept thinking of how, in the bush, Rocky and I used to race up to the top of the red ridge like Stallone running up the steps. At first I could remember so real how lekker it was to look down on the world with Rocky by my side, but after a while my lungs forgot whatzit like to breathe free. There comes a point when only the shape of being a human is what's keeping you walking and working and eating instead of dissolving like air, cos you've got nothing inside. The misery becomes comfortable like tattered old pyjamas you prefer to lekker new clothes.

If there hadn't been that accident I would have kept on wearing my misery to hide the empty shell I was already. But my bru, this is what happened. As I was working underground one day, I felt my heart go boomboom and those scars I got from the thorns that night started hurting. I looked down and I swear to God, they were bleeding fresh after all these years. I dropped my tools and came running out, not minding the okes asking where I'm going. The second I came up to the surface I felt the trembling of the earth and the kukuku sound of rocks crumbling like a piece of stale bread. I was the only one out of hundreds of okes who got out alive.

As soon as I could, I went to see Rocky. The scaly look on the director's face showed me something was wrong before he even said anything. Rocky hurt a visitor, he said. He picked up a boy with his trunk and threw him down. Fortunately, the boy only broke his leg.

There must be some kinda misunderstanding—Rocky would never hurt anyone, I said. Let me see him.

Rocky was standing alone in his kraal, chained in place by each of his ankles. He become massive, his head was the size of a sofa and his tusks were like the masts of a sailboat. He become an elephant that could touch the sky, if only he were outside. I went up to him and he rumbled low, moving side to side. It's me, Rocky, my little boet, I said to him, slowly getting closer. I told you I'll come get you. Slowly I got near him until I was close enough to lay my hand behind his ear, just where he liked being pet. He sighed and shuddered, he put his trunk around my shoulder. That's when I saw the bladdy crack on his trunk and a nail sticking out of it. What they do to you, I cried. My tears fell into his wound as I dug the nail out with my fingers.

Even when I showed him the nail, the director said he had no choice but to put Rocky down, it was the law against human-hurting animals. I said, Give me just a week and straightaway went to Da, who asked round and told me to go see this manager at P___ National Park.

Driving there, I liked how it was full of trees, much greener than where I grew up. Soft gray clouds were draped over the mountains like laundry out to dry. There was plenty of food for Rocky as far as the eye could see. The manager—a red-faced oke who had a habit of licking his lips—agreed that adding one elephant would trouble his park none, it had enough hectares to feed a thousand elephants and one or two more wouldn't change much. He assured me that culling wasn't going to be done and besides, he would be sure not to eliminate Rocky as Da's special china and all. Rocky would probably get to sire his own calves and help diversify the park's gene pool. The only thing was that the translocation fee was expensive, 40,000 rand—as he said this he had this look on his face like I wouldn't have that kinda moola. I pulled out an envelope, counted out exactly 40,000 rand in cash, and pushed it across his desk. It was all the money I got from the mining company after fifteen years' work plus the accident, my bru. On the way out I was feeling how Stallone must've felt at the end of fifteen rounds with Creed, beat up but happy like I done everything I could.

Just before the park exit, I stopped to go to the toilet and have a sarmie at the picnic table. By now the clouds had disappeared and the sun was shining through, and it had turned into a lekker mild afternoon, all green and gold. There was only one other person there, a ranger, and I offered him half my sarmie cos I was feeling thankful. He took the sarmie and said, Thanks, bru, and we got talking. I asked if he liked working at the park and he said, Ja, it pains me tho that so many animals get poached now, my bru. What do you mean? I asked. I have to be working all the time, bru, like I can't take a single day off. It's been six weeks since I seen my kids, he said. If it's a full moon, we lose one. If I'm off-duty, we lose one. The truth is, bru, it's an inside job. The manager opens the gates to whoever pays him enough.

I went back to the sanctuary and told them what happened. An under-the-table elephant like Rocky, with those huge tusks of his, would be killed off no problem on the next full moon, likely sooner. I had no more money to offer them, could they please keep Rocky alive sumhow? The director shook his head, saying even if it weren't against the law, they be forced to shut the whole place down if Rocky hurt another visitor.

The last time I saw Rocky, he was still in his chains. He vacuumed up the corn kernels I held out, then sniffed me all over my face, neck, and hands looking for more. The wound on his trunk had all healed up like his body was determined to keep going and that made me angry too—why do we even bother if it's all going to end the same way no matter what? What the point? I just couldn't lie to him and say, This isn't the end—I'll come get you. He saw me crying and he knew what it was, he just gently wrapped me up in his trunk and pulled me close to him. My bru, that elephant wanted to console me even while knowing he was ganna die.

Afterwards, I went to Durban and got on a fishing boat. I wanted to go far far away from the bush. So I went all over the world for a long time—Indonesia, Australia, Japan, Ecuador. I never known that there be so much blue, it flooded and covered all the dry earth inside me so I forget.

Just once there was a call for me when I was at the sea, on a clear, windy night. A lawyer saying Da drove his Jeep into the bush and was found days later, picked clean to the bone. No rifle or pistol on him. The house was the park's, the lawyer said. He could sell whatever's inside and send me the money. I said, Burn everything—but then I looked up and saw these same stars I seen that first night with Rocky. We were crossing over the equator just then, thousands of miles of black sea allaround, black sky up above. I said, No—bury them. Before I heard him reply, the wind picked up again and the signal died.

On the boat the okes kept to themselves, it was the kinda place where you didn't ask alotta questions. Only the wind and the waves talked to me until I could understand them. It was in Alaska while watching out for the icebergs that I heard a voice say, I want to go home. *Home!* It gave a little shout. I heard it as clear as you can hear my voice right now.

It took me months to come back to KwaZulu-Natal, to the park where I grew up. Eventually the sea ended and the smells changed, all that heat and dust and sky dry and high like it's about to shatter. The closer I got the more my heart swelled up inside, making my bones ache to hold it in. The sun was scorching the earth, and the trees had mostly gone brown in the heat except for the acacias. I saw that they grown stronger from drinking in the blad that they drew. The house with its kraal came into view, and I couldn't help myself—I stopped the car and jumped out. I fell knee first into the red earth and curled into a ball, rolling around and gathering up the dust in my hands.

I knew then why everything was. The voice I heard wasn't mine, it was Rocky's. Some part of him had been inside my body all along, that's why I never jumped off the ship even tho the thought crossed my mind about a thousand times. When you really love someone, a little piece of you goes inside them and a little piece of them goes inside you—it's most often your ma and da but it doesn't have to be your family or even human, my bru. They won't ever die as long as you live, understand. And if I had to say one truth about my life bru, it's this. Out of everything and everyone on earth, I loved that elephant the most.

A gentle rumble came from up above just then. I looked to the top of the ridge and saw him holding up the blue sky on his shoulders, waving his ears at me like a butterfly.

BIOARK

It is 99 degrees and sunny. Beyond the length of my legs glowing palely on the deck, the Atlantic Ocean is glistening like a pile of rubies. I close my eyes, relishing the sensation of my body dissociating from my mind. If you focus your attention enough, you can perceive the heat dissolving the walls that keep you intact. As I'm beginning to feel my thoughts flapping in the air around me like flags, someone sits down on the lounge chair next to mine. Says, "Hi Herman."

"Oh, hi—Esme," I say, pausing between each word, to let her know I'm not keen on conversation. She notices my reluctance but carries on with undiminished enthusiasm. She decided long ago that we had to be friends because we're the two Argentinians on board.

"It's dangerous lying out like this. Are you even wearing sunblock?" she says, pointing at her long-sleeve top. "I always cover up everywhere, but I still got melanoma on my chest. Luckily, the doctor found it in time ..."

"I'm sorry to hear about your melanoma," I say. Esme is nice, but she's earnest. She hunches, and she wears her graying hair in a ponytail. My eyes float past her to the top deck, where a woman wearing a yellow maillot is leaning her tanned forearms on the railing, gazing out at the wine-dark ocean. The Ark's CEO. Flanking her on either side are her grandchildren—the twins, Apollo and Diana, both long-limbed and dazzling. Not one person out of the three looks a day over twenty-five. Esme follows my gaze and regards the trio with scorn.

"Do you remember what life was like before the Ark, Herman?" Esme asks.

"You know I've never been on land, Esme. I was born here," I say curtly.

"I was only very little, but I remember I had a grandmother. A grandmother who didn't look the same age as her grandchildren." Esme gestures a little self-righteously at the top deck.

"It doesn't bother me," I say, closing my eyes again. "It's like the heat, Esme. I don't like being hot, but if you suspend judgment about something that can't be helped, it is most bearable. I'm guessing there's a point at which it may even become pleasurable. And that might be the only thing we can count on. Do you know what I mean?"

I open my eyes to a sliver and see that the chair next to me is empty. I lean back once more, feeling the gentle unraveling of my judgment and my lover's gaze peering down at me from the top deck. It warms my skin like a second sun.

Nothing further disturbs my sunbathing until the doctor's appointment at 4 p.m. When Dr. Evgenia sees my inflamed skin, she does a body scan and then hands me a pill. I gulp it down with a small shot of water.

"How are you feeling otherwise? Physically? Mentally?

Emotionally?" she says, tilting her heart-shaped face to one side. I allow myself to imagine that she really cares—that after all these years, we might even be something like friends. It makes me say things I wasn't planning on saying.

"To tell you the truth, I feel mixed up lately. A bit *jumbled*," I broach.

"What do you mean by that, exactly?" Dr. Evgenia says clinically, but she leans forward with a slightly predatory twinkle in her eye.

"Like, I get up in the morning and don't realize where I am for a minute. Sometimes I even forget my name. I feel disconnected. Is this because I'm getting old?" I drag my hands down my face, blinking a few times. Dr. Evgenia laughs like a string of pearls breaking loose on the floor.

"Nonsense, you're still young. It's probably heat exhaustion. Stay out of the sun and keep hydrating, Herman," she says, binding my upper arm in rubber. She plunges her needle in and the glass jar fills with my blood.

Maybe because I *am* suffering from heat exhaustion, I don't have much of an appetite at dinner. I draw my knife along the side of a roasted tomato and watch the slick red liquid spread over the plate, drenching the potatoes and the chicken piccata. My eyebrows knit together reflexively—I don't like my foods mixing on my plate. But I suspend judgment until the dislike faintly resembles a preference, and something akin to relief washes over me. And that transition point is what thrills me. I really feel as though I've discovered something important—though it's not what I was originally seeking. I spent so much time looking for the answer and have recently come to accept that my query was always futile.

For years, every evening after dinner I used to cloister myself with books in the virtual library. It's not a densely populated area of the Ark, compared to the swimming pool, the Orgy Dome, or the Fertility Temple. These places are all IRL—after the virtual versions of these spawned *their* own virtual places, the top deck made the rare unilateral decision to shut down all the virtuals except the library, which never had an IRL version to begin with. Most of the directives are designed to promote physical activity, as it is healthy and humanizing. Also, people get irritated easily due to the heat, so it's necessary to keep them constantly amused— and reading is generally considered to be too depressing.

Personally, I like the solitude and the musty smell of books; but tonight I head down to Euphoria, one of the Ark's main dance clubs. As I descend the mirrored stairs, the rattling of the bass jolts me awake. The bodies thrusting and shaking to the music appear blue in the black light. The tiny pricks of strobe light cause them to resemble human cutouts made of the night sky. I mix in with them, letting the music take hold of my body.

While dancing, I sense someone nearing me like a comet. A hand wraps around my waist, a shoulder bumps into my back, and then a soft mouth presses into mine. When we pull apart, Apollo smiles and places a little unmarked pill between his lips. I suck off the pill with a kiss, and my body instantly floods with the aroma of pineapples. Every caress feels like diamonds—from Apollo to my left, and from Diana to my right. I'm in the middle of the two sides of the same coin, and I try to discern the difference between them. They look and taste so similar, I can only tell them apart by the hardness of Apollo's angular body and the softness of Diana's. When they kiss each other, it is like a solar eclipse, a perfect cancellation that amounts to spectacular nothingness.

I don't know how we make it back to Apollo's suite. Next thing I know, I'm in his bed alone. I turn my head toward the bathroom and see him standing there in a towel, outlined by the soft orange light. Silhouetting best captures the classical perfection of his form, I've found.

"Was brushing my teeth," he says. "Go back to sleep, Herman." It's the unexpected gentleness in his voice that catches me off guard.

"Come here, I want to tell you something." I motion to him. "Please. It's important."

Apollo turns off the bathroom light and climbs into bed next to me. Starts stroking my hair off my face.

"Have you ever wondered if we could have prevented all this from taking place?" My voice trembles a little and I hope he doesn't notice.

"What?" He smiles, but the corners of his mouth droop. "I thought you were having fun."

"I don't mean us. I mean the earth is a toxic wasteland, the oceans red and dead with algae—no trees, no animals, just some humans left in this Bioark and some Biodomes," I say with a lump in my throat, but Apollo laughs.

"You're still high. One with the universe and all that." He chuckles, stroking my cheek with a finger.

"No, I'm serious. You've never wondered if there was a point in time when all this could have been prevented—when we could back away from the cliff, instead of jumping off?" I catch his hand in mine and squeeze it for emphasis.

"Honestly—" Apollo stares at me solemnly with his purple eyes. "No. I haven't."

"Well, I have. I read every book I could get my hands on, I did my research. Because time doesn't just run—it *spreads*. Everything that ever happens is like a piece of a puzzle, and if you take out one piece, all the rest can be undone."

"But time does flow in one direction, Herman," Apollo says indulgently. "Yesterday, today, tomorrow. In that order."

"Think about photons. Or think about how electrons orbiting a nucleus don't pass through different levels but jump from one quantum state to the next. That proves not everything is linear. Time could also be quantized into chronons—puzzle pieces. Do you see?"

He doesn't see but wishes to humor me.

"So you found the puzzle piece when our predecessors irrevocably destroyed the planet?"

I nod. Apollo looks genuinely surprised, maybe even impressed.

"But I have no way of taking out that puzzle piece, and it's worse than if I'd never looked for an answer." I notice that he doesn't ask when that point of no return was. He is not curious, so I don't offer.

"I feel like someone who wanted to know how deep a well was, and so jumped in it."

"Herman, this is what I find so attractive about you," he says. "You're so passionate. But you're looking at it the wrong way. The world hasn't been destroyed. The Ark keeps many people alive and happy. Education, freedom, healthcare for all, and no meaningless labor, poverty, hunger, racism, or war. I think you should focus on the positive. Stay present, yes?" He kisses my neck, and my body relents even as my mind stays unconvinced.

"You would never have been born if things hadn't happened the way they have. We would never have met. And for those two reasons at least, I wouldn't want anything to have been different," Apollo says. I feel my eyes water, and my throat crams with the words. But we've never said I love you before, and I don't want to ruin the moment by getting greedy. I stay quiet.

"Instead of reading, maybe you should take up something else that's a little more uplifting, Herman. Yoga, sophrology, or sculpture. I could come with you," he continues.

"I'd like that," I croak. My eyes keep watering, but Apollo pretends not to notice. This is one of the many reasons I feel for him—his perfect balance of warmth and coolness. It almost hurts to contain the four-letter word that bubbles up inside me. I fall asleep in his arms, thinking that maybe being together with him is a worthy trade-off for the lost world.

****

I wake up when the sun rises and fills the suite with peaches-and-cream light. Outside the window, the sky and the sea are aflame in shades of crimson. The alarm clock on the dresser blinks as it changes from 07:39 to 07:40 a.m. It is December 22.

Apollo is already out of bed—from the splashes in the bathroom, I surmise that he's taking a shower. Soon he comes out, a towel wrapped tightly around the sharp, V-shaped valleys of his pelvis. Apollo's beauty is such that even his standing with his damp skin by the nightstand looks like poetry.

"Good morning," I say to him.

"Hey, how'd you sleep?" he says with a smile, but I realize in that instant that last night's tenderness has already disappeared.

"I slept well, and you?" My voice sounds timid. Apollo is focused on pouring himself a glass of water and opening a bottle of meds. He senses me staring and turns around.

"Oh, this," he says, showing me the little white pill in his palm. It is almost identical to other drugs the passengers of the Ark take, but with a tiny red *E*. It's why Apollo and I look so different, although we were born in the same year. When I've lost all trace of youth, he will continue to be his most ravishing age for as long as he desires.

"Who gets this besides you?" I ask.

"Grandmother, Grandfather, my mother, Diana." He counts them in his hand. "Can you imagine if others got a hold of this?" He scoffs; then he checks himself, noticing my changed expression. "I don't mean you, obviously."

I start rooting around the bed for my clothes, nonetheless. He doesn't try to stop me—another warm day is stretching out ahead of us, and neither of us are in the mood to linger anymore. But when I'm dressed, he reaches toward me in a conciliatory way—and the pill gleams on his palm like a pearl.

"Here, try it," he says. "It's not going to make any difference to me if I skip one dose."

"And if you skip one week? Or a month?" I pick up the pill with two fingers. Instead of swallowing it with the glass of water on the nightstand, I put it in my pocket.

"Then I guess I would die, like everyone else," he says, giving me a quick kiss on the cheek. "I'll see you at yoga."

When I step outside, something seems to be missing from the usual morning cacophony. I don't realize what it is until the speakers and the screens announce that the Ark has temporarily stalled due to technical issues. There is no mention of whether the solar panels, wind turbines, desalination filters, or the engine have malfunctioned. Making matters worse is the fact that we're not stranded in the relatively mild waters of the Arctic, but right off the Strait of Gibraltar, where Europe kisses Africa. Without movement, the temperature on the deck shoots up to 105 degrees.

At 4 p.m., I retreat gratefully to Dr. Evgenia's office with its blast of air-conditioning. She asks me the usual questions, hands me my pill, and draws my blood efficiently—so efficiently indeed, I can feel the color draining from my face.

"You don't feel mixed up today?" she asks, pulling the needle out of my arm.

"No, I'm great," I say, and she flashes an approving smile.

As a matter of fact—I decide while sitting down to dinner—I do feel better than I have in a long time. What a difference it makes to have just one person to whom your existence really matters. That is reason enough for being. Fuck history, fuck wishing things were otherwise. I vow to get my dosha balanced, stay present, and enjoy my time on the Ark.

On my plate is a palm-size slab of steak. I drag my knife along the meat, unleashing its pink juice dotted with yellow bubbles. It coats the green peas and the potatoes in a film of bloody grease. I saw off a piece and pop it in my mouth, rolling it around slowly. Apollo was right about advances in humanity—progress *has* been made. Take this meat, for example: not so long ago, people used to raise animals in cages, so cramped and filthy that they were driven mad to cannibalism. People fed chicken shit and feathers to cows, pig parts to pigs—really anything to pigs—and animals asphyxiated from their own gases in lightless warehouses. Imagine the trillions of bodies bred only for suffering, and the Final Extinction makes sense; we had to be wiped clean to start fresh. Those animals are long gone, but this cultured steak is cellularly identical to a cow—without any pain or death. In the lab, it will continue to multiply as long as it's fed the correct combination of sea salt and platelet-rich plasma, which is full of growth factors. A daily donation is a small price to pay for a lifetime of sustenance.

After dinner, I go back out to the deck for sunset yoga class. I usually try to avoid these, as it's Esme's favorite activity, but this is the first time Apollo has suggested doing something together as a couple, and I feel a minor happiness.

The sun hisses into the ocean like a piece of hot iron, but it's still unbearably hot. There is hardly any wind and the Ark hasn't moved an inch all day, three miles off the Rock of Gibraltar. I look around in vain for Apollo's golden head above the others, and give up when the teacher starts asking us to call attention to our heart chakra. That's when I realize that Esme is also missing from her usual spot in the center front. In fact, I haven't seen her since yesterday afternoon.

I try to concentrate on warrior pose but keep straightening my legs and popping up out of the lunge. Sweat keeps dripping down into and from my eyes like I'm a crying mess. I don't even like yoga—I was only doing this for him, and so of course he's a no-show. It's always the same story with him, there's always some excuse. Something like, "I thought we were going to talk about it again? I didn't know it was a set thing." Why I keep falling for this bullshit is a mystery even to me. I leave yoga angrier than ever—the others could probably sense the poison emanating from my savasana.

Once I'm in my own room, the coolness of the bed begins to calm me down. I chug a tall glass of icy water and instantly feel much better. I strip off my drenched clothes and lie on top of the bed. The sheets are like yogurt on my sunburned skin and I turn off all thoughts.

* **

I wake up when the sun rises and fills the suite with peaches-and-cream light. Outside the window, the sky and the sea are aflame in shades of crimson. The alarm clock on the dresser blinks as it changes from 07:39 to 07:40 a.m. It is December 22.

Diana is already out of bed—from the splashes in the bathroom, I surmise that she's taking a shower. In fact, it was the sound of the water that woke me up from a dream. I try to remember what it was I saw. It was something immensely large and white. As I comb my brain for the word, Diana walks out in a towel. Without looking at me, she faces the full-length mirror with an absolutely straight back, unwraps the towel from her body, and uses it to squeeze her wet hair dry. She always does it this way—no hunching to fuss over the big towel or a smaller towel, no delicately hiding her body or putting it to its best advantage, no wild shaking and rubbing of the hair. She deploys zero extra movements when she comes out of the shower. I find this both glorious and a tad frightening.

"I had a dream last night where I saw a bird," I say to her. She turns to face me. She drops the towel; her short, damp hair curls around her face, leaving the rest of her body uncovered.

"A what?" she asks. "A . . . bird?"

"Yes, it's a kind of animal that flies."

"Oh," she mutters, turning back. "*In*-teresting." She opens her closet and starts pulling on her clothes in the same, streamlined way. She stashes her screen in her pant pocket and walks toward me—to kiss me good morning, I expect, but instead she reaches for the glass of water on her nightstand and swigs it to wash down a little white pill. It's similar to mine, except it's labeled with a red *E*.

"Apollo just messaged me. There's some problem with one of the propellers. Looks like we've been grounded since early this morning," she says. "We're not going to be moving for a while—maybe weeks. Maybe months."

"Well, it will get fixed eventually," I offer. She stares at me with her unblinking purple eyes.

"Herman, why do you think we need to be moving? Nothing is waiting for us anywhere. But we need the propeller to be working at all times because then it feels as though we're going somewhere. Without movement, everyone would go mad within days."

I know this, of course. Diana leaves before I can apologize for being so insensitive. I pull on my clothes and lumber onto the deck, where people are already looking restless and trapped, gazing out longingly at the horizon where the Rock of Gibraltar is stabbing the sky. To its left, Europe shimmers red, as if it were on fire; to its right, Africa does the same thing. I shade my eyes with my hand, straining to see any movement on land.

When the temperature rises to 105 degrees, I turn around to look for Esme. Before the project was abandoned as too resource-intensive and the animals were depopulated, Esme was the species keeper of the Ark. I want to ask her about the bird in my dream—but I can't find her anywhere.

It is in the virtual library's twenty-first-century collection that I find a documentary of this bird. It glides with its enormous wings, its feet barely skimming the waves. A disembodied voice narrates:

The wandering albatross can grow to a wingspan of twelve feet, the largest of any bird. By using a special soaring technique, after takeoff it can fly six hundred and twenty miles in a day without once flapping its wings. In a single two-week trip out to forage for its young, an albatross parent travels five thousand miles out to its fishing spot across the ocean, and flies back another five thousand miles. An albatross chick is entirely dependent on its parents until its fledging—but once the time comes, the parents don't return to the nest. In the colony, the abandoned juveniles must starve until they are light enough and desperate enough to attempt flight. On its very first try, the juvenile makes for

the ocean and there it stays for several years, only return-
ing to land when it is ready to breed. Indeed, an albatross is
born to fly. It is the only animal known to circumnavigate
Earth . . .

Because of the documentary, I am late to my appointment
with Dr. Evgenia. When I open the door, I'm surprised to discover
that today she looks almost identical to Diana. I arrange my face
to appear unfazed and answer only the questions she asks.

"How are you feeling?" she asks, handing me my usual pill. I
take it in my mouth and sip the shot of water.

"Good," I say through my teeth, and attempt to smile naturally.

"Not mixed up today?" she asks again with flashing eyes, and
I try to remember if they were always purple.

"No," I mumble. "Doctor, I was wondering if we could skip
the donation today."

"Why? What for?" She lowers her voice. "You just said you're
feeling well."

"I mean, is this really necessary?" I blurt out, unable to hold
any longer. She knits her eyebrows together sternly.

"If by necessary you mean essential for all the food produc-
tion on the Ark, then yes," she says, binding the rubber around
my upper arm.

"This is what keeps us all alive, Herman. There is no other
way. And when you suspend judgment of circumstances over
which you have no control, it begins to feel a lot better."

She plunges the needle into my arm and the glass jar fills with
my blood.

I decide to skip dinner. After coming back to my room, I spit
out the white pill she gave me, which I kept tucked under my

tongue. My bed, desk, chair, and lamp are all familiar to me, they look like they're saying hello to me in their inanimate language. I strip off my clothes and lie on top of the sheets, hearing the buzz of the water dispenser until even that quiets down and the world becomes mercifully silent.

***

I wake up when the sun rises and fills my room with peaches-and-cream light. Outside the window, the sky and the sea are aflame in shades of crimson. The alarm clock on the dresser blinks as it changes from 07:39 to 07:40 a.m. It is December 22.

I lie on the bed for a moment, feeling the weight of my body coalesce around the part of me that is me. My nucleus, my center of gravity. It takes a while to feel it. I recall that my name is Herman. I am thirty-six years old. It is December 22, 2090, and the Ark has been sailing for 19,578 days. At the beginning, it was hoped that the wildfires, floods, and droughts would eventually stabilize after purging much of humanity, and the Ark would deliver the last inhabitants of Earth to land. That hope has long been abandoned, so now we live for only the world of the Ark. It has its rules, and it demands to be obeyed. My unwillingness to accept it as my world and my life has drawn its notice. It offers me my desires, tells me what to think, and waits for me to submit. If even a single one of its offerings were real, I would have capitulated a long time ago. But something tells me that my awareness of the games of the Ark has broken open its farce.

On the nightstand, my screen chimes with a message. I'm being summoned to the top deck. As I put the screen in my pant pocket, something small and round brushes against my fingertips. I take it out: it's a little white pill labeled with an *E*. I've never seen it before, nor do I know how it got there, but instinctively I realize what it is. A chronon that has slipped through the cracks. I put the pill back in my pocket.

From my second-class cabin, I have to climb quite a bit to get to the staterooms at the top. I pass by the hydroponic garden, where the chef is harvesting fresh vegetables for tonight's dinner. Then there is the Olympic-size swimming pool of pristine, turquoise water, where people are cooling off from the heat. It is already 105 degrees, and sweat streams down between my shoulder blades as the speakers announce that the Ark has temporarily stalled due to technical issues. I go up to the middle deck, where they host yoga classes; finally, I reach the top deck, and the imposing stateroom overlooking the entire Ark.

I knock. A woman's voice says, "Come in."

I open the door and walk inside, and the blast of air-conditioning envelops my skin like a salve. The room's walls and floor are covered entirely in soft, ivory cashmere save for the floor-to-ceiling windows on two sides. A slender golden woman is looking down at the decks; when I enter, she pivots around on her tan, high-arched feet to face me. The Ark's CEO.

"Madame E," I say, and she giggles.

"Call me Stella. I feel like we've known each other long enough, Herman . . . Please sit down." She pours two long-stemmed glasses of water and hands me one.

"You know, I still remember what champagne tastes like. It's incredibly hard to engineer. Much more so than a tomato, for example. One day, though . . ."

"Why did you call me up here?" I ask, putting the glass down on a table. She sits on the couch and takes off her heels, crossing one beautiful leg over the other into an upside-down V.

"I heard you were having trouble keeping straight. Your

health, mentally and physically, is spiraling. Evgenia told me. Also
Apollo. He's so fond of you," she says, smiling. I don't smile back.

"Apollo came to me to talk about this, and we had a chat as
a family as well, which we always do when it comes to issues as
important as these. All of us agreed at the end." She opens a black
marble box on top of the table. "We want you to have this."

She takes out something small and extends it to me. Through
the glass of the bottle, I see tiny white pills labeled with an *E*. I
take the cylinder in my palm and turn it slowly front to back.

"Why do you want to live forever?" I ask.

"How ghastly," she balks. "I don't want to live forever—it's
just to stay young so I can enjoy my life while it lasts . . . And yes,
it will be a lot longer than for most people. But I'm hardly im-
mortal."

"Well? What's the point of being young when this"—I point
to the windows, where the ocean glimmers a diabolical red—"this
is what we have left?"

"You misunderstand me." She sighs. "If by aging, I could
bring things back, then maybe that's a fair critique. But it would
accomplish nothing, would it not? Those are two separate issues,
are they not? There's nothing wrong with living a full life."

"You go ahead," I say, putting the bottle down on the table. I
dig out the single pill from my pocket and put that on the table
too. She looks displeased for the first time.

"Listen, Herman," she says, uncrossing her legs.

"You think that once upon a time, humanity still had the
chance to set this right, before passing the point of no return. Because you think, if only people had known . . . But let me tell you
something, as someone who was alive through it all." She takes
the bottle from the table and unscrews the cap.

"Oh, people knew when the point of no return was. They

knew how to fix things too. Scientists issued cris de coeur from Durban to Oslo—it was in the newspapers every day, all that breast-beating exhortation. It's not because they didn't know, Herman. It's because they didn't *want to*." She puts a pill on her tongue and takes a sip of water.

"Between a mild inconvenience now and an intense suffering five years down the road, people always, always choose the latter. You see, the vast majority of humanity can't plan beyond one year of their own future, let alone five years of collective future. Speak in terms of ten years and people don't think it's any closer to them than a black hole millions of light-years away. It's fascinating to me, this failure of imagination that defines *Homo sapiens*! My husband and I were aware of this when we built the Ark. At the time, it was supposed that things could still be remedied, everyone talked as though something must and will be done, but we knew otherwise."

She smiles. Her ninety-year-old face is exquisite, the same way very young people's cheekbones seem to exude their own light. After just one dose, the growth factors have already taken their lap around her body.

"I regret that this will sound self-congratulatory, but we've managed to advance humanity even under these challenging circumstances. No one is left to die of hunger, poverty, disease, or violence on our Ark. Things are much darker on the Russian Bioark—Evgenia tells me every week how glad she is to have defected! And the world that you think was robbed from you—that world was a selfish, doomed place. I assure you."

I get up quickly, feeling dizzy for a reason that has nothing to do with dehydration.

"If that's all you had to say, I'm leaving." I knock my knees against the coffee table in my haste to get out.

She says mockingly to the back of my head, "Really, I expected more from you. Denial is so unoriginal . . ." before I close the door behind me.

It dawns on me that she's right, even if I don't like to believe it. As I run down the staircase, the sun is burning high above the Rock of Gibraltar and drying the red ocean drop by drop. The Ark hasn't moved an inch since this morning and there is no breeze on the deck. My shirt is drenched by the time I reach the second-class cabins and stop in front of Esme's door. She's the only one I know who still remembers life before the Ark—the oldest person here besides Madame E. I want her to tell me something different about the way the world ended. Stories about her grandmother for example, who tended a garden and planted red flowers for the hummingbirds and purple ones for the bees; the fountains in Buenos Aires; at the beach, her father teaching her to swim by backing away from her, saying "I'm right here! I'm right here!" while she kicked furiously, crying; how undaunted and self-sacrificing people tried until their last day to stop the Extinction. How when the Ark first sailed, the oceans were still blue and full of fish.

I knock and call her name.

There is no answer. After a few times, I push the door—and am surprised when it opens. Her furniture is exactly the same as mine, only neater and exuding a certain Esme-like quality. I don't know how anyone has the patience to tuck in their sheets under the mattress every morning, but that says to me that this is Esme's room, and I feel comforted. Although I should turn away, I walk inside, innocently as though just checking out the view from her window. That's when I notice a piece of paper folded neatly in half and lying on top of the bed.

I reason that no one would leave a piece of paper like that on their bed for themselves—it feels meant for me. I pick it up and open it. It's a line drawing of a bird with enormous outstretched wings. Its long, sturdy neck leads up to a small head and a gently hooked beak. I don't know what bird this is and have no idea why Esme would leave something like this for me—if it was meant for me at all.

I drop the drawing and drink from her water dispenser; I can't remember the last time I had water, which must be contributing to this feeling of disarray. The moment it courses through my body I feel instantly calmer.

As I consider my options, my screen chimes with a message from Apollo: *where are you? want to do dinner together—just the two of us?* I almost reply to him but stop myself because doing so will make me forget any other thoughts besides those about our relationship. My screen says it's 2 p.m., which means I have two hours until my doctor's appointment. I decide to spend that time looking for Esme.

I go around searching for her in all the usual places, but she's nowhere to be found. People don't even seem to know who I am talking about. I feel as though I'm making her up until the yoga teacher says he missed Esme at yesterday's evening class. Which means I was the last person to see her yesterday afternoon on the lower deck. I recall that it was 99 degrees and sunny, and the Atlantic Ocean was glistening like a pile of rubies.

I go back to my room, log in to the Ark's virtual library, and begin rummaging through the pile of books and films. In a dusty corner of the early-twenty-first-century collection I find a documentary and tap to press Play.

The wandering albatross can grow to a wingspan of twelve feet, the largest of any bird, the narrator intones as the bird in Esme's drawing floats effortlessly over the ocean.

By the time I finish the documentary, it's already 6 p.m. I've missed my doctor's appointment, which is the only requirement on the Ark. But what they may or may not do no longer frightens me.

When I reach the lower deck, the huge, bloated sun looks like it's about to ram into Earth like a meteor. All is red—the sky, the sun, the sea, the Ark, and even my arms and legs. There's no one outside; everyone has fled to the safety of air-conditioning. I don't even bother looking at the thermometer—it's hot enough that I can feel my organs boiling.

But I keep squinting into the sinking sun, because there is a small dot weaving gracefully between the fanned-out rays. A bird with impossibly long wings. It's been circumnavigating Earth looking for a safe place, I'm sure of it. All this time. What is a day? It's only the time it takes Earth to orbit on its axis, and when you're a bird that can fly six hundred miles without once flapping, that's meaningless.

Like all truly beautiful and real things, the albatross shines briefly before vanishing completely.

I could go back to Apollo and wake up tomorrow morning, and the one after, and ever after, growing younger and happier. I do believe we will one day tell each other I love you, and mean it. I already do mean it.

I could jump off the ship, and if I don't die from algae poisoning or exhaustion—both highly likely—I may be able to swim three miles to the Rock of Gibraltar and die there. Or I may live to see another day, meet Esme, and perhaps other survivors. What makes up my mind is the fact that what happens to me is less important than what it does to everything else around me. By the very air I breathe, I break and remake the world. That is what it means to be alive.

For the first time today, there is a strong gust of wind that whips my hair and stirs up waves. I feel ready. I will dash across the empty deck and shoot out from the gunwale with my arms spread wide, willing them to grow like the wings of a young bird. My body will arc upward, far longer than I imagined possible, and for a moment I will believe I won't ever come down, I will fly.

But the wind stops blowing and I catch myself at the gunwale. My screen is chiming and again it is Apollo, begging me to come to dinner with him. *Where are you?* he asks with a petulant neediness that delights me, then follows it up with: *I think I'm falling for you.* The fragility evident in this text makes me forget why I was so desperate to leave the Ark in the first place. After all, I am not a bird—I am only human.

OLDER SISTER

The phone rang a little after midnight. It was Mother, crying hysterically that she couldn't get a hold of Older Sister. *You have to go find her,* she sobbed. I hadn't seen or heard from Older Sister in years. But I knew the way to her house, which wasn't far from my place. *I'm leaving right now,* I told Mother, grabbing my car keys and stuffing my feet into my shoes all at once.

Before I had any understanding of who I was, I knew what I was not: Older Sister. Before the incident of the ghost, we really didn't have anything in common that confirmed that we were sisters. But that came much later. Let us start from the beginning.

Mother says Older Sister didn't hate me when I was born— that also came much later. When Mother gave the newborn me in a white onesie to Older Sister to hold, I was already almost half her size at age four. (Mother would always add a side note that she fainted while giving birth to me.) I was a large, slobbery monster-in-making, for the moment as helpless as a dragon egg. Instead of rejecting me or "accidentally" dropping me on my head, Older Sister cradled me in her little lap and gave me a bottle. I gobbled the formula down and gurgled happily into her face. She smiled, thinking that she was finally getting a friend, or at least some kind of a pet.

Very soon after, Older Sister was put in charge of taking care of me. Mother and Father both worked at our downstairs grocery store from 8 a.m. until 9 p.m., seven days a week. Like most other parents in Koreatown, they left us alone without any hesitation. Usually Mother came up to our apartment on the second floor to cook us lunch and dinner, but by the time Older Sister was six or seven, she could heat up leftovers in the microwave or use the footstool to stand over the stove and cook ramyun. She was incredibly precocious and full of gumption. She could fold any kind of irregularly shaped laundry with her tiny hands. Due to her neatness, ladylike demeanor, and Lilliputian stature, she was nicknamed Little Princess.

I, on the other hand, gradually began to overtake her in size but not in smarts. I had a notably weak subconscious connection to my bladder, meaning I wet my bed regularly until I was five. I also would get dry sinuses and have trouble breathing, which I tried to help along by excavating my crusted nostrils, which in turn led to frequent nosebleeds. Mother, probably unaware of what I was doing, lamented that I was healthy everywhere except for my respiratory system.

Despite my bed-wetting and other unpromising traits, Mother insisted I was naturally smarter than Older Sister, al-though Older Sister worked harder. She said this right in front of both of us, several times throughout our growing up. I was able to count to a hundred before my first birthday; and I taught myself simple arithmetic at age two, when Older Sister entered school and carelessly left around her workbooks of which I then took advantage. I'd seen and heard her pronounce things aloud from the symbols on the paper, and it all clicked together in my brain. From then on, I read every book our parents gave Older Sister, which she herself didn't read. Our parents thought this presaged some brilliance on my part. Older Sister thought this presaged something else a lot more sinister. Now I look back and wish Mother wasn't so vocal about an opinion that is, after all, an opinion—neither truth nor fact.

Maybe this was one of the reasons Older Sister started bullying me. She asked me to do a lot of things—fetch the remote, turn on the lights, put the dishes in the sink—and anytime I refused, we would get into a huge fight. Even though she was little, her pinches and kicks could sting pretty badly, so naturally I couldn't take that lying down. Mostly though, I had to listen and obey. This was because we were Korean. Both Mother and Father punished me if I addressed Older Sister by "you" or her given name. If she asked me to do something within reasonable limits, meaning humanly possible, I had to do it. That was my role as Younger Sister. And her role as Older Sister was to take care of me. It was as though they had given birth to us so we'd become each other's servants.

I eventually settled into my role, because after all, she took good care of me. We still had our fights and differences, of course, but we were very close by the time of the incident of the ghost.

It was Thanksgiving weekend in my senior year of high school, and Older Sister was visiting home—she was a senior at UC Berkeley at the time. After she'd gone to college, our parents and I had moved to a two-bedroom apartment, so she always slept in my bedroom during breaks. We were tucked in my full-size bed, all the lights turned off, chatting about the boys we had crushes on. Then midgiggling, both of us stopped and gripped each other's hand.

"There's someone there," I whispered, pointing at where my open door led to the hallway. Someone's breathing had suddenly materialized at that exact, empty space—similar to wind rushing through the chimney vent but distinctly, rhythmically human. And in fact, male. It was a cold and dreadful presence unlike anyone I knew.

Older Sister whispered back, "I know! We have to wake Appa up!"

For the next several minutes, we screamed: "Appa! Appa! Someone broke in! Wake up!" We screamed until our voices became hoarse, but there was zero sound from our parents' room. The intruder's breathing continued all the while. Finally, as the bigger and stronger of the two of us since birth, I said:

"I'm going to jump out of bed and get a golf club. Once we have that, we can turn on the lamp." Father's golf bag was near the bed, and I was frightened of facing the intruder in full light without any weapons. Older Sister agreed this was a sound plan. So I scrambled out of the bed, retrieved a golf club from about five feet away, and immediately leaped back to the inviolable safety of the mattress. For a brief moment I felt very proud of my willingness to swing an iron rod at an intruder if it was to protect my Older Sister. We counted to three, then she just as bravely reached out her arm and turned on the lamp on the nightstand.

There was no one there.

"I heard the breathing until a second ago! Nonstop, since the moment we both said there was someone there!" I said, and Older Sister looked at me in astonishment.

"What? I didn't hear anything!" she said.

"What do you mean? You said there was someone there."

"I *saw* someone—it was an older man, wearing a bathrobe . . ."

We ran to our parents' room and woke them up shrieking as though we were little children. They claimed they didn't hear our screams from my room—how that's even possible in an 800-square-foot apartment, I don't understand. All of us firmly believed that Older Sister and I had witnessed a ghost. It was one thing if a daughter encountered paranormal activity, but two daughters? Now, you couldn't have a shred of doubt. Older Sister was, after all, a biology major and a woman of science. She wouldn't have believed in ghosts without seeing one with her own eyes. For some reason, the fact that the ghost showed itself to Older Sister and me at the same time, but in different formats, made me think that we really had to be related by blood. We now had something that we shared only with each other, that neither of us would ever forget.

Skeptics would say that we'd already undergone a life-changing event that predisposed us to imagine something like the ghost. To explain, I'd have to start with our parents.

Mother and Father came to LA in 1975, ten years after the repeal of the Asian exclusion laws legalized immigration from Korea to the United States. Mother had been a high school math teacher in Korea. Like Older Sister, she was petite with large, chocolate-colored eyes. She did everything very neatly and gracefully, even when she was restocking shelves. When she sang hymns at Our Lady of the Grotto Korean Church, head covered in a diaphanous veil strung with fake pearls, I thought she looked exactly like the Virgin Mary.

Father dutifully attended mass every Sunday, but he just stood there with his hands pressed together and didn't even go through with the pretense of lip-syncing. The only time he sang was when he got very drunk on soju and started belting out "Real Man," a military anthem that every able-bodied man in Korea knew by heart after three years of compulsory service. Every Korean wife also knew the lyrics in full, as it's something like a trumpet heralding the return of an inebriated husband. It went like this:

> *Born a man, there are many things you*
> *   must do*
> *But you and I, we lived for the glory of*
> *   guarding our country.*
> *Comrade-in-arms! Our friendship was*
> *   forged in the crucible of battle.*
> *Between sunset and sunrise over the hills*
> *Our parents and siblings sleep soundly,*
> *   trusting me.*

Father had served in the marines between his freshman and sophomore year at Seoul National University. So by the time he graduated with a degree in physics, he was already twenty-six years old. He found a position as a particle physicist at a government research institute. But he quit three years later—there the family lore gets a bit messy, something to do with his controlling Older Brother. Maybe he just got tired of being Older Brother's lifetime indentured servant. Instead of telling Older Brother to mind his own business, Father flew across the Pacific with two hundred dollars in cash and all his worldly possessions in a single carry-on suitcase. Since he was a physicist back home, he had nothing else to do here besides work at a grocery store. That's where he met Mother. An exceptionally strong and tall man for a Korean of his generation, Father courted her by stoically lifting all the heavy objects in the store. There was no barrel of soy sauce, no cardboard box of Napa cabbage, no eighty-kilo bag of white rice that he wouldn't move around for her. She expressed her gratitude by cooking for him a humble yet deceptively complex stew—because it was made of homemade kimchi, which no bachelor would ever dream of attempting on his own. Their courtship was by nature practical but filled with gestures that said, *I will take care of you.* Even from the day they met.

This isn't to suggest that they were always passionately in love. Our parents fought as much as Older Sister and I did. It usually began with Mother saying something sarcastic and biting, and Father grimly drawing his lips into a tight line and seething and snorting like an angry bull. Then Mother would lavish him with the saltiest, dirtiest profanities that one couldn't imagine pouring out of such an elegant mouth. After about fifteen minutes of this, Father would explode, and switch to the formal speech, bellowing something like, "PRAY, YOU SHOULD NOT ADDRESS ME THAT WAY!" Which was how you knew he was really furious.

I want to clarify that this didn't mean they had a dysfunctional marriage. Along with the age-enforced hierarchy, another fundamental aspect of Korean culture is arguing easily and intensely. Sometimes these distinct aspects of our culture overlapped, as when two strangers got into a fight (for example, over a minor car accident) and things quickly devolved into a shouting match over who was the elder. Then they would be at each other's throats, trying to prove their age, whipping out their driver's licenses, since the one who had seniority immediately secured higher ground.

The thing about Koreans is that we are a fighting people; our most common cheer is literally "Fighting!" pronounced hwa-i-ting, with a defiant shake of a fist. But this is because we've had violence thrust onto us for centuries, and we've learned that to not fight back is to accept being destroyed—to consent to being erased forever in history. And no, I'm not just talking about the Japanese occupation (although, good one!). People who think erasure has ceased after World War II, at least in *developed countries*, are those who have never had to fight for their existence.

It was April 29, 1992—I was five and Older Sister must have been nine. We were back home from school; Older Sister was doing homework, and I was reading and waiting for her to finish so she could make us dinner. Mother came rushing into the apartment, and without saying hello to us, turned on the TV. On the local news, there were people breaking windows and hauling out bottles of liquor and electronics. The overlaid text in blue and white said that the defendants were acquitted of the beating of someone named Rodney King. The police in riot gear were trying to make arrests and then retreating as they got rocks and bottles thrown at them. People were shooting guns, turning over cars and setting them on fire. A white truck driver passing through South Central stopped at a red light, was pulled out of his car by a crowd, and beaten on the head with a fire extinguisher right in front of the camera. The screen returned to the footage of liquor stores, grocery stores, swap meets, and wig stores rising up in flames and spewing pillars of black smoke into the sky. Looters were streaming out of them, carrying everything from stereos to diapers. Many of the stores were owned by our acquaintances from church or around town. The news didn't say anything about what happened to those Koreans.

Mother ran back downstairs to bolt the store with Father, and then they came upstairs and turned on Radio Korea. The Korean owners in South Central were on air, saying that the police weren't responding to their 911 calls. Everything they owned was looted, destroyed. Someone whose wife was shot and injured said that the police car just drove past the rioters firing at his family. The radio host urged people to stay and fight for their property. If they didn't stand up for themselves, no one would protect them, their families, and entire life savings. They called for volunteers from other parts of California to come to our aid. As the night deepened, the gangs began distributing leaflets in South Central with the words "Koreatown is next." At the same time, a Korean gun shop owner was handing out firearms to anyone who didn't have one already. College students still in their teens arrived from UC campuses to help the *ajussis* and *ajummas*. The older men with military experience all knew how to shoot guns, and ex–Special Forces and ex-Marines came forward to lead groups in the worst-attacked blocks.

With other men, Father took a handgun to the rooftop and started watching over our building. Before daybreak, the rioters had reached us. They shot at Father and the *ajussis*. He shot at the sky. They got closer, shooting at him again. He shot at the sky again. The third time they shot at him, he aimed and shot at them, to show he could. They threw something at our store, shattering the glass. From the roof, he could see the tongues of flame licking the edges of the storefront. He lost his mind thinking about us and Mother on the second floor.

He flew down to the first floor and put out the fire with an extinguisher while the other Korean owners in neighboring rooftops provided him with cover. Mother screamed, "*Yeobo!*" grabbed a bucket of water and ran downstairs as well but returned when Father yelled at her to stay with us. She went back to calling 911 every five minutes; no one answered. We knew through Radio Korea that the LAPD and the National Guard were blocking off the road between Koreatown and Beverly Hills, not between South Central and Koreatown. Black smoke bloomed upstairs even with the doors and windows shut.

Of all this, what I remember the most is how unimaginably loud a gunshot is, how you flatten yourself on the ground and instinctively hide from windows even without Mother's orders. I crouched down and Older Sister covered me with her tiny body, like I was the turtle and she was the shell.

I don't understand how Father could remain standing while being shot at, let alone have the wits to shoot back, for three nights and three days. That's how long he stood guard over us without any sleep. When they set fire to the store for the third time, he didn't have any strength left to put it out before everything we owned was gone.

After the rioters finally retreated, and the blackened sky over LA faded to gray, Father didn't even check his burned and stolen inventory—he came upstairs, closed his eyes, and instantly fell asleep. Mother took Older Sister with her downstairs to assess the damage and see if they could salvage anything, because even though she was a child, she was an Older Sister. There wasn't much they could do, however. Not a single roll of toilet paper was left. It wasn't just us—almost everyone we knew lost their livelihood.

Maybe this was why we were always looking over our shoulder, because anyone could take everything away from us in an instant. After something like this, it's actually a miracle we didn't imagine an intruder every single night. Which is one reason I think that the ghost was not imaginary. He only ever showed himself that one time.

Like most of the other Korean businesses, our store hadn't been insured. Our parents didn't say much about what happened. They never sang anymore, although at mass Mother knelt and sobbed under the cover of her white veil.

During this time, Older Sister and I came to a mutual understanding that we must never fight again in front of our parents; their survival depended on it. We had to be well-behaved, obedient, academically outstanding. To give them something to live for. When Older Sister told me to fetch water, I fetched water. When she ordered me to brush my hair, I brushed my hair. I walked around outside picking up lost pennies from the ground, collected them in a pouch, and gifted them to Older Sister. She started to look upon me favorably as a needy but loyal pet.

This pattern kept up as we progressed through grade school, middle school, and then high school. We were both excellent students, because it was what kept our parents alive. It was as though we were feeding them through IV, our accomplishments in lieu of sugar water. Older Sister stayed up until late at night, copying things from the textbook in her ant-size handwriting, sometimes breaking into tears if she couldn't understand a theorem.

But I confess, I didn't even have to pay attention to class; most of the time I was drawing pictures on the margins of my notebooks. My favorite doodling subject was a dream house—I loved imagining a huge, marble-floored mansion with a courtyard. There would be a black stone pool with water lilies and flame-colored goldfish that came up to the surface when I walked near them. Light-filled rooms with French doors opening out to the terrace, where we could all eat under the curtain of wisteria on summer nights. Lots of different types of trees and flowers on our vast, emerald estate. While drawing, I kept one ear open to the lecture so that when the exams arrived, I could pull out the information from the back of my brain. Taking my SATs in junior spring, I was internally humming Tchaikovsky Symphony no. 5 and admiring the notes flashing before my eyes in color; I was more taxed by filling out the bubbles than by figuring out the answers. It was as though someone was whispering "A, C, E, B . . ." into my ear for three hours.

After I finished, Older Sister, who was home for spring break, came to pick me up.

"How did you do?" she asked as I clambered onto the passenger seat. Next to my bigger frame, it was Older Sister clutching the wheel who looked like a student driver.

"I don't know. I think I got everything right," I said.

She drove on without saying anything for a while. Four years previously, she hadn't done as well on her SATs as she'd hoped. She didn't get into her first choice of Stanford and had settled, tearfully, on UC Berkeley. I turned on the radio and started humming along to that Fastball song about an eternal summer and a road that's paved in gold. The sun was shining in slats between the palm trees. Then the balls of light became gradually larger until they burst like bubbles, washing the world in pink.

When we got on Wilshire Boulevard, she said to me, "Good job." We drove the rest of the way home in silence.

A few weeks later I opened the envelope from the College Board with shaking hands. It was 1600, a perfect score. I called our parents at the laundromat, which they'd opened several years back with some miraculously scavenged funds. They did an extraordinary thing in my honor: they hung up the CLOSED sign and came home early for the first time in their lives.

But it was too early to celebrate. We knew that I wouldn't be able to attend college without a full-ride scholarship. Our parents had trouble enough sending Older Sister to a state school. They'd poured every possible line of credit into their laundromat, which sometimes didn't take home any money at all. Still, I was optimistic—I had the perfect score, the perfect grades. In the fall of my senior year, I applied to Harvard, Yale, Princeton, Dartmouth, and a fleet of well-regarded non-Ivies. I didn't waste any application fees on schools that could be considered safety by any normal standards. Also, out of solidarity after Stanford wounded Older Sister's pride, I resolutely refused to show them any interest.

In spring, one by one, the rejection letters started to arrive.

When I first got the telltale thin envelope, I pored at it, incredulous, then said to our parents, "Well, maybe the next one." They clapped my back and repeated what I just said. But after the third or fourth letter, even Father struggled to string together faux-hearty phrases. Mother clasped my hand in hers and recited the rosary. Maybe because of that, a couple of schools then offered admission with a partial scholarship, not even close to what I'd need.

Older Sister came back home for spring break. We were lying in my bed with the lights turned off, like when the ghost had appeared. She was thinner than ever from studying too hard—she was going to be taking her MCAT soon. We lay in silence, flattened from all the fighting we had to do to keep surviving.

"Do you have any more schools you're waiting to hear from?" she asked.

"Only Princeton," I mumbled. "I didn't expect to be rejected so thoroughly."

I could hear Older Sister contemplating. Then she said, "Well, they think we're just test-taking machines and pianists. Number crunchers. Not creative."

The thought had crossed my mind; even my college guidance counselor had said something similar to my face. All the Korean kids are the same, he'd said casually, as though by that he meant to point me to the path of light. Kindly helping me realize how undesirable I was.

"How much did NYU offer?" Older Sister asked, and I told her.

"Then go to NYU. If worse comes to worst, I'll find a job after graduating and pay your tuition."

She said this in her Older Sister voice, the way she ordered me to wipe the dining table with a wet paper towel so she could serve dinner. She was directing me so she could take care of me.

"You know what color heart you have?" I asked her, and she shrugged in the darkness.

"It's gold," I said.

"Can you see it?" she asked.

"Yes. Right in your chest. Not cold like a crown, but warm like honey and wheat."

After Older Sister went back to campus, I knelt at the foot of my bed and really, truly prayed for the first time in my life. I prayed that she would ace her MCAT and get accepted to the best med schools in the country. I cried asking that Princeton would accept me with a full scholarship so that Older Sister wouldn't sacrifice her life for me. If only one of us could succeed, however, I prayed that it would be her.

A few weeks later, it was Father who stopped the postman as he was filling the mailboxes of our apartment building. Usually the postman got testy when people lurked around open mailboxes, but he saw the desperation in Father's face and fished around in his bags for one big envelope from Princeton addressed to me. When Father saw how stuffed it was, he couldn't contain his excitement and opened it himself. Reading the welcome letter, Father collapsed on the spot and wept. He hadn't even done that when the store burned down.

Despite the exorbitant cost of doing so, our parents made the decision to drop me off at Princeton together. I remember that when we passed through FitzRandolph Gate, a dozen or so blond kids were playing croquet on the lawn in all-white ensembles. They did that every year to intimidate the freshmen. With more experience I would have said to our parents (loudly so the croquet players could hear) "Don't mind those asses—they still think this is Antebellum South," but at that moment the best thing I could do was hold my head up high and walk past them with a stolid expression. Our parents did not seem to mind. They were in awe of everything.

Older Sister did not come with us. She wasn't accepted to any med schools, and she'd herself just begun the physician assistant program at OHSU in Portland. I'd prayed that if only one of us could succeed, God would pick her over me. I didn't think it was possible to say this to her without sounding condescending. When we spoke on the phone, our voices were stilted, a little unnatural. But when I told her about my three lacrosse-player suitemates (Maggie, Caity, Frannie) who threw a rager at our place without asking the other suitemates' permission, covering every inch of common areas in red Solo cups, and then barfed and peed in their own rooms, all within the first week of freshman year, she softened.

"White girls," she said, shaking her head on the other side of the phone.

"I didn't know girls could be so disgusting," I said. "By senior year, I swear someone will have pooped on the floor."

I already knew then that these girls without any intelligence, integrity, or basic personal hygiene would reap the benefits of college more than I ever would. The Maggies, Caitys, and Frannies attracted their similars like magnets while looking right through me as if I were invisible. White people liked to say "all Asians are the same." They didn't seem so bothered that some white people were indistinguishable from one another, in their manner of speech, the expression of their eyes, hairstyles, Longchamp tote bags, sport coats, sailing shoes, and pearl studs—and yet *their* sameness was an advantage, and their power grew the more identical they were.

Because I didn't fit in well in college, Older Sister forgave me. This isn't to say I didn't love Princeton: the jasmine vines climbing up the dorms, the dear old gargoyles, the libraries smelling so wonderfully of books and knowledge, the neo-Gothic arches ringing with a cappella groups, the sun spreading its orange over the boathouse. Everything was so comely, like a modern American setting for a Jane Austen novel. I loved college the way I loved *Pride and Prejudice*—without seeing myself inside it, without expecting to be loved back. I liked sharing these reflections with Older Sister; it seemed to make our relationship better.

After I graduated, I started working as a reporter at a local paper in Inyo County. My beats were environment and social justice, although the reality of the job meant I covered everything from county fairs to black bear antics in people's backyards. I was earning a starvation wage, but I tried to give each story a certain literary flair. And for every third missing-cat-finds-owner story, I could pitch something a bit more ambitious.

I was happy—but our parents were busy hiding their disappointment. They'd thought I'd be less head-in-the-clouds after college. After three years as a reporter, Mother said to me, "When you run too fast around the track, you find yourself actually *behind* other people—that's what you're like." Sometime after the paper folded and I lost that job, Father pleaded, "We know you're brilliant. You'll be successful one day. But we're getting tired of waiting. So tired."

"I'm freelancing, Appa. I have a feature article coming up next month," I told him. I was proud of that story, about how droughts

on the West Coast were affecting Indian reservations. "I just want
to make a difference."

"You could've made a difference in another way. Become an
environmental lawyer. Or worked for the UN."

I faced similar questions from Older Sister when I visited her
new house in the Southwest Hills of Portland. She'd met her husband while working at OHSU; Bradley was a cardiac surgeon, as
well as a grandson of a self-described "old Portland family" that
had made its fortune in logging about a century ago. (I find de-
scribing the generationally wealthy as "old ___ family" is just as
repugnant as the word "expat" to describe rich foreigners. All families are old—we *are* the same human race.) I had to pass through
two sets of wrought-iron gates to reach Older Sister—one for the
community and one for their home. The latter one reminded me
especially of FitzRandolph Gate. The house itself was not floored
in marble, nor did it have a black stone pond filled with goldfish
that came up to the surface at the sound of my footsteps. It was
much less Metropolitan Museum of Art than that, and much
more Spanish Colonial, with a sparkling turquoise swimming
pool surrounded by potted olive trees. But it did feature French
doors opening out to the deck; in the evening, you could sit there
and see the lights from all the millionaire mansions embedded in
the surrounding hills, like a chandelier of dream houses. Older
Sister put me in a guest room with a four-poster bed; she knew I
would like that. But before dinner, when Bradley excused himself
to play *Call of Duty*, Older Sister sat me down with iced coffee and
began grilling me.

"How's the work situation?" she asked. To my family, my career was never just "work" but a "situation."

I told her about the drought story. She pressed me about the
fee, which I admitted to her was probably less than what they

spent on the champagne for their birdbath every month. Then she told me to stop joking around.

"You're thirty-one and still barely getting by. When will you grow up?" she said, adding milk to her coffee and stirring just a little too hard.

"Says someone whose husband is playing video games at forty-five!" I said.

"Leave Bradley out of it. It's the only thing that relaxes him after surgery," she hissed.

"I'm happy. Honestly I don't care about money," I said, leaving out my views on capitalism.

"Listen, if you really want to make a difference, you have to have money. Do you remember the riots? Why didn't the police protect us?"

Of course I remembered the riots—I thought of it often, and whenever there was a sound of glass breaking, an alarm, fireworks, or even gleeful laughter outside. Some nights, I lay awake, and the image of Rodney King flashed before my eyes, pleading before the cameras, *Can we, can we all get along? Can we stop making it horrible for the old people and the kids? I mean we've got enough smog in Los Angeles let alone to deal with setting these fires and things . . . It's just not right. And, just, uh, I love—I'm neutral. I love every—I love people of color. It's just not right. It's just not right, because those people will never go home to their families again . . . Let's, you know, let's try to work it out.*

"Because we were Korean. They only protected the whites." I paused; and then because Older Sister was looking at me intently, I continued.

"White people had been oppressing Black people for centuries before Koreans arrived here, and then they encouraged a narrative of Black-versus-Korean fight—better for them to keep the minorities fighting against one another."

Older Sister drew a deep breath as if I were truly dense. "You were only five. What do you really remember about all this?"

I snorted in indignation, but the damage was done: like a lettuce seed tossed on top of soil, doubt took hold with the greatest ease. Didn't I smell the black smoke, crouching under the gunshots right beside her? I shivered; she carried on with an air of exhausted patience.

"We were attacked because we were poor and powerless. Money and power go hand in hand. If you want to help poor people or animals or whatever else you care about, fine! Save the planet, won't you? But do it after you take care of yourself first—look at you, *you're* the thing that needs saving! Otherwise, you're just a blip in the sea of chaos."

I sat in silence, watching my glass leave a ring of condensation on the coaster.

"I got you vegetable skewers and portobello caps," Older Sister said, rising from the couch. "Bradley will grill them first so they don't touch the meat."

It was a brutally hot day. The leaves of trees had curled up and fallen like pencil shavings on the second-floor deck, where Bradley tended the fire, an additional burst of heat inside the bigger heat.

"Isn't it too hot? Maybe we should just cook inside," I said to him, passing a platter of skewers.

"Oh no, we don't. We're American, and this is the Fourth of freaking July." Bradley laughed and took a swig of his dripping scotch soda. A vast ring of sweat was spreading across the back of his polo shirt, and the scalp under his thinning crown was turning salmon pink. I fanned myself using my hand, moving the steamy air as much as an ant's breath.

"You know, I grew up in these hills and the heat wave gets worse every year," he said more soberly, perhaps reconsidering his declaration of citizenship and trying to find common ground.

"Portland used to be in the seventies, with some days in the eighties, all summer. No one had air-conditioning. Now we'll get a few weeks over a hundred degrees, some days over a hundred and ten degrees. And we have a whole fire season—like Southern California! It's tragic."

"I know. I lie in bed thinking about these things," I said, eyeing Older Sister who had come out to swap the finished skewers with raw steak. "Have you seen that viral video of the starving polar bear chewing on car seats? I cried for days." Even as I recollected, my heart came up to my throat; I coughed to clear away the tears before they happened.

"Stop being so dramatic. Of course polar bears will go extinct," Older Sister said as Bradley plopped down the bloody steak on the grill. "Or they'll interbreed with brown bears and create a new hybrid species. That's just how evolution works. It's nothing to get emotional about."

My vision was raining red. I gripped the edge of the teak table for balance, but everything was red and hot to the touch.

"Sweetheart," Bradley said to Older Sister in his peacemaking voice. "Can you please get me another drink?"

"You know what I mean. She gets overemotional, almost hysterical, about the wrong things," Older Sister continued, as if I weren't there. I stormed back into the air-conditioned kitchen, but not before hearing her say, "We used to think she's high-functioning…"

I had planned on staying with them for a week but gave some excuse and left after three days. Older Sister only saw me off at the front door, not even bothering to come out to the driveway like we're supposed to do. But when I called our parents afterward, they took her side as usual.

"Don't fight with Older Sister," Father said to me on the phone. "She's older than you."

"Appa, what about you and Uncle? Sometimes you have no choice but to stand up to older siblings," I said. He expelled a sigh, and I could sense his shoulders curling forward. All the years of working at the grocery and the laundromat had shrunk his height by at least two inches.

"When we lost everything in *sa-i-gu*," he said, meaning April 29. "It was your uncle who gave me the money to start over. He did yell at me that I should've done something a bit safer, somehow. He suggested a laundromat, since other people's clothes are not really worth looting, and washing machines can't be stolen in a shopping cart. I took his advice and his money. He never asked for a cent back."

Soon after the call with Father, I decided that I was in a rut and that something had to change. My work was freelance, and I wasn't gaining new opportunities by staying in the same sleepy county where the only local paper had gone out of business. And I'd had enough of California, its heartlessness and forgetfulness. A land of the Lotus Eaters. When my lease ended, I packed up my things in my Hyundai Elantra and drove up north to start over in Portland. I took Highway 101 rather than the faster I-5, as it seemed more celebratory for new beginnings. The road hugged the cliffs, where the blue-green ocean crashed with all its lunar force and turned to foam brighter than the purest snow. Redwoods older than Christ stood like cathedrals, with bark instead of limestone and needles instead of steeples. Farther north in Yachats, I saw a gray whale and her calf on their way up to the Arctic. In this place where the forest met the sea, the air tasted sweet and mossy. Mother was right—I must have had respiratory problems, because I inhaled as though I was learning to truly breathe for the first time. I pulled the car over and prayed that I would be nothing more than a tree or a whale or a wave, things that stay what they are and do not presume to become something else. Anything that remains true to itself is beautiful, which is why nothing in nature is ugly when you look at it with the right eyes.

I was fully myself, aware, and happy those three days, more than at any other time of my life before or since.

I settled into a one-bedroom apartment in Goose Hollow, an old and leafy part of town not far from Southwest Hills. I dusted the walls and the five-inch space above the kitchen cabinets. I put up frames, hung curtains, and rolled out rugs. I bought and fretted over potted plants. I was ready to show—not just my parents and my sister, but especially myself—that I deserved my little corner of the universe. But months flew by and I didn't get anything more than a handful of first-round interviews. The general feedback was that I was either too experienced or not experienced enough, and was a "strong writer" but not quite a "culture fit." Whenever I heard this, I stood in front of the mirror and tried to pick out the wrongness in me: Was it my glasses, thick, uncolored black hair, large frame, or my overenthusiasm for certain things? Would I be a "culture fit" if I were tiny and beautiful like Older Sister—or would that still not help, since she was also Korean?

I continued pitching freelance articles and picking up copywriting gigs, but only a trickle; one of my existing clients had budget cuts, and another hired an in-house writer. On a particularly bad day, when I found out only an hour after an interview that they went with another candidate, my Elantra puttered to a stop on the side of the road. I was still in my interview clothes, just desperately trying to get home so I could experience misery in more comfortable attire. I curled up behind the wheel, pressed in by walls of blackness, until a kind passerby jump-started the battery for me.

When I finally made it back three hours later, I called Older Sister, partly because of what Father told me about his own Older

Brother. And because before she got married, she used to say she loved me.

Lest I change my mind, I told her that I was in trouble financially. Older Sister said, "All right, I'll help you."

"Really?" I was bewildered; I didn't know it was going to be so easy. "Thank you."

"Yes. Now listen carefully," she continued. "Take $5,000 of your savings and invest exactly as I tell you. It's going to be a combination of crypto and gold. You're going to 3x your money in a year, maybe less." She said this like "three ex," not "three times" or "triple."

"You're going to tell me how to invest?"

"Yes. And just trust me—I 10xed my investments in the past five years. The market's crazy—made a hundred k this week. So you're in good hands."

I didn't have $1,000 in my checking account, let alone $5,000. I told her as much, mixing in liberal amounts of swear words that I'd never, ever been allowed to say to her. And that was when we last spoke on the phone.

Since then, I'd gotten her updates through Mother. In due time, she and Bradley separated. He hired one of the most expensive divorce lawyers in town (his birthright!), but in the end, Older Sister and *her* team got most of what she'd wanted, including the house. It was easy to prove that she had contributed more to their combined wealth through her stellar portfolio, and besides, it was Bradley's alcoholism that had caused their marriage to break. After that, Older Sister quit working at the hospital and turned to investing as her main line of work. When Father was diagnosed with stage I lung cancer, caused by all the years of working with dry-cleaning solvents, Older Sister found him the best oncologist in LA and paid for his entire treatment. I moved back home for a year to help take care of him with Mother, and never once ran into Older Sister. When he went into remission, she sent our parents on a cruise to Mexico. They drove up to stay at her Southwest Hills house once every six months. But even after I came back to Portland, I never joined them.

Earlier this week, a wildfire started in Forest Park. It was only one of two dozen fires simultaneously raging in Oregon, and it was thought to be minor. At one point, it was believed to be fully contained by a circle intentionally burned by firefighters, who then moved on to other areas where thousands of acres were being swept under the Great Flood of flames. It's unclear what happened next—maybe a stray lightning, wind, or a tossed cigarette butt—but somehow the fire came roaring back to Forest Park. Within an hour, it pressed down toward Southwest Hills, burning everything in its path.

I was watching the live news: aerial shots of cars driving through the smoke, trees wrapped in orange like pillars of fire, a house catching flame and then blackening like a marshmallow. It was a Spanish Colonial with a familiar wrought-iron gate at the end of its long driveway.

I felt seasick and tasted bitter orange at the back of my tongue. That's when the phone rang and Mother's voice pleaded with me to find Older Sister. I grabbed a bottle of water and ran out, pushing my arms through my sleeves at the same time.

On the drive up to her house, all of our arguments and happy memories passed before my eyes like strips of film. We were each other's servants, polar opposites, best friends, worst enemies, and finally, strangers. I honestly didn't know if I loved her anymore.

But there were people you had to save even if you didn't love them, maybe even if you hated them—and Older Sister was on top of that list. I kept one hand on the wheel and called her with the other hand; it immediately went to her voicemail.

As I rounded the corner to her neighborhood, the bloodshot moon was rising over the crackling hills. The smoke thickened and I rolled up my windows reflexively. I passed a car speeding down the slope in the opposite direction. I froze, wondering if this was foolish—Older Sister likely had already escaped, and I was jumping straight into a wildfire. It would have made more sense to stay at my apartment and call 911. As I was panicking, the community gate came into view. For a moment I thought it would be locked and that I'd be forced to turn back. But maybe because of the fire, the gate had been left open. There were trees on either side of the road that were catching fire, screeching as their branches broke off. I pressed the gas and drove forward.

From afar, I could catch glimpses of stately houses in flames. No people. But a golden retriever ran away before I could pull over and get it inside my car. When I reached her house, her wrought-iron gates were locked. I jumped out of the car and retched as the smoke filled into my lungs. I poured my water bottle all over my shirt and a bandana, wrapped the latter over my face, and squeezed between the gate and the wall. Up ahead, the house was creaking and shuddering in the center of a fire as bright as the sun. Even from some fifty feet away, the air was so hot that my shirt was drying almost instantly. I didn't think Older Sister could be alive if she'd stayed inside; she had to have escaped a long time ago. But I couldn't leave without trying to find her.

"Unni!" My shout was swallowed by the house's groans. A huge bough fell off a tree in front of me, making an explosive sound like a gunshot. When I stopped cowering and raised my head, I saw her standing barefoot on her second-floor deck. In a nightgown and a bathrobe.

I ran to her and shouted at her to jump down. It was about twelve feet from the deck to the ground, and she hesitated for a moment before the advancing fire made up her mind. She dropped straight like a needle into the bushes below. When I reached her, she was curled up in a ball and her right lower leg was jutting out at a strange angle. Falling out of her bathrobe pockets were flash drives and bars of gold that, even in the split second I glanced down, gleamed hypnotically under the dancing light. I scooped her up—she weighed nothing in my arms—and ran. But before I'd gone twenty steps, I saw through the gates that my Elantra was on fire. It screamed as its hood exploded, shooting a plume of flame to the sky.

"The swimming pool," Older Sister coughed out. I was fading from the smoke, but my feet still obeyed her. I laid her down in the pool—she gripped onto the ledge—and I draped a tarpaulin over the water before crawling underneath it.

"Keep breathing through your nose," Older Sister hissed. "Close your mouth . . . Stay calm."

I don't know how long I was floating underneath the tarpaulin. I went into a strange dream state where I thought we were in another world, waiting to be born. In this dream, she said, "Come, we have to go," and I refused—I was too tired. "Fine, I'll go first and then you follow," she said. "Keep to the west, and then to the east. I'll meet you there."

I opened my eyes and saw her pale blue face by the light filtering through the tarpaulin. The noises outside had died down. I threw off the cover and gulped the gray air. A mourning dove's call was rising clearly through the ashes. It sounded searching, like a question. Perhaps the dove too was looking for its family.

I laid a hand on my sister's chest, feeling for warmth from her heart. It was still beating, but its color had faded like the morning moon. I needed to make her stay.

"Do you remember the incident of the ghost?" I whispered. I thought I saw a trace of a smile on her lips. I knew what that smile meant. Would it surprise you to know that this was one of our favorite memories, perhaps our happiest moment? How even when we were screaming, we had gripped our hands together with joy? Because what is life but a series of ghosts we face alone, with no one to tell who would understand? When living like atoms with infinite spaces between one another, understanding was rarer and better than love. We believed each other once. We were the lucky ones.

MOUNTAIN, ISLAND

There are certain expectations of drama before meeting seven of the world's most stupendously adored men on the planet. But when I arrive ten minutes early at the subterranean bar of the Mercer Hotel, the members of KBB are already seated around a U-shaped banquette, laughing unselfconsciously over drinks. All seven of them—J-Raw, Min, Bin, Be, Hyuk, Eddie, and X, and not a handler in sight. They could pass for a group of NYU sophomores—albeit uncommonly well-dressed and fabulous Korean exchange students with a penchant for Gucci—if not for the unmistakable aura of those who find global success at a very young age. It's the lethal focus of a sniper couched in self-determination, wrapped up in a Rick Owens cardigan of I-know-this-could-all-end-tomorrow-and-that's-okay.

"It might sound strange, but we didn't plan on world domination, you know?" Min, the leader of the group, says thoughtfully between sips of his coupe. When he notices me staring, he immediately slides the glass across the table and urges me to try it.

"It's a Concord grape rum punch with an essence of coconut milk." He smiles, and two dime-size dimples appear on his otherwise perfectly smooth skin, like the season's first snow covering a frozen lake.

I take a sip—it's a ridiculous-sounding drink that actually tastes really good, which reminds me so uncannily of KBB's music. Their latest single, "Juicy Melon Bomb," is a sugary pop paean that could turn a triathlete into a diabetic from just one listen. It hit one billion views on YouTube in mere sixty-five days and has since soared past the two-and-a-half-billion mark. Translation: a third of humanity have danced and sung to KBB's latest creation, a fruity riddle ("I wanna, I wanna / Banana, guava, papaya / I gotta, d'you wanna?") that doesn't feel like it has anything to do with groceries.

When I press them about the true meaning of "Juicy Melon Bomb," Bin clears his throat. "From the moment we heard this track, we just felt so uplifted," he says, cross-shaped earrings dangling from his lobes.

"We thought, if this song could make our fans happier by even 5 percent, who are we to not release it into the world? We don't have the right," Be finishes Bin's thought. (*Be* means *rain* in Korean, but also *to be* in English: "It's always been important to me to show our fans that staying true to yourself, to simply *be*, is life's greatest achievement.")

There is an island where the Indian and the Pacific Oceans meet. Its name is homophonous with "flying boat ax moon," but it doesn't actually mean anything. The only people who know the name are the locals, and they themselves just refer to it as "Island." Although it's roughly twice as big as Iceland, no one outside the region can remember where it is or what it looks like on a map. There are too many islands and archipelagos in this part of the world, and this particular island has not done anything to stand out among the rest. It used to be a beastly green jungle ringing ceaselessly with birdsong, encircled by a turquoise shallow sea. Just like every other island.

Now, though, only the old have any memories of the sea. The young know only the Mountain, a landfill that can comfortably fit two hundred Olympic stadiums within it. The islanders live, work, get married, and go to school there among the trash. When Agus was born in his family's shack, the trash outside used to tower just above their roof. Now in his final year of primary school, his front door directly opens out to a shuddering, rotting mound soaring sixteen stories into the air. With his hands firmly around his backpack straps, Agus picks his way to school, the vertical hills rising around him like the world's smelliest library shelves. *Squelch, squelch.* The path is mostly clear, but sometimes he has no choice but to climb over a mound. He keeps his eyes on his sandaled feet, automatically avoiding the ballooning plastic bags of decaying fruit, dog feces, and takeout boxes stuffed with leftovers and greasy utensils. When he notices something that might be of value—a pair of soiled jeans, a full bottle of shampoo, a magazine—he stashes them quickly in his backpack.

By the time Agus makes it to school, the kids are arrayed around the Teacher under the tarpaulin roof. She is the most successful person the Mountain has ever produced. The Teacher attended a high school outside the landfill and then won a scholarship to go to a university overseas, in a polite country that once, regrettably, colonized the island. Besides her, no one Agus knows has ever left the Mountain. She teaches the children everything, from multiplication to brushing teeth, everything they need to know in the world. When they shout the right answer as a group, her eyes twinkle and her face turns red with too much excitement. But when she thinks they've all left, she puts her head in her arms over her desk, like a nesting bird.

At noon, Agus eats his lunch with Heru, who offers some rice and an avocado pit with plenty of brown flesh still around it. Agus splits his melted chocolate bar and a very brown banana. They've eaten lunch this way for the past six years, always sharing everything. Afterward, Agus shows Heru his morning findings.

"Nice jeans?" Agus says, hoping to swap them for a pair of sneakers Heru found a week ago and is now wearing. There is a finger-long gash running through the center of each sole, but they're Nikes, in Agus's size (and too big for Heru). The swoosh logo shimmers holographically in the morning haze.

"Nah." Heru shrugs. "I could use the magazine, though."

"Just take it," Agus says, irritated. He vows to find a better pair of Nike shoes, even if it takes him a year.

Agus gets up to chuck his banana peel into the nearest mound from their open-air classroom. That was the one thing the Teacher could not do—make them use a wastebasket. There is no point in politely gathering up rubbish when you live in a landfill. Every day the trash mound inches closer to the school, and one day in the not-so-distant future it will subsume the little shack like a festering monster. This is what eventually happens to everything and everyone at the Mountain.

As Agus trudges back toward Heru, a sudden noise makes him jump and look up. A violently noisome wind starts blowing all the plastic bags around so that the children scream and pull down on the corners of the classroom's tarpaulin roof. A helicopter is descending right onto the mattress-size schoolyard, which is free of rubbish. A suited man, the Director of the island's Tourism Bureau, emerges from the aircraft, plugging his nose with a finger and looking rather green in the face. He ignores the questioning dark eyes of the children and walks straight to the Teacher. The helicopter's rotor blades meanwhile still revolve, threatening to blow away the roof (and the clinging children with it) and hiding the man's words to the Teacher in the ensuing noise.

". . . It's absolutely depressing. You can feel the annihilation of the soul at a place like this," the Director is saying, when the blades finally quiet down. "You have to show them hope, positivity."

"Yes, Director," the Teacher agrees dutifully.

"It's important to affirm happiness, human feelings. No one would fly halfway across the world and pay money for a tour to find out that they're bad people," the Director continues. "The fact that they would *choose* to be here when they could be anywhere else, shows that they're decent, upstanding. People are good, good are people."

The Teacher's mouth quivers as her eyes scan the cordillera of brown detritus that's surrounding them on all sides. Here's something you don't learn in school: every color mixed together doesn't look black—it's brown. And that's the true color of humanity; every object that has ever passed through hands, every bodily function, and actual body parts (most often clumps of hair but also occasionally, a foot, a hand) end up forming a piece of a great brown stain that is the evidence of human existence. It's visible from space. Its ugliness tears through your sinuses, rots your skin, and shatters the thinking and feeling parts of you.

"I believe we're born good, sir," the Teacher says, and the Director barks out a laugh.

"Positivity! I like it. Just remember this one thing." The Director whispers something into her ear and climbs back into the helicopter, which shreds the putrid air until it is aloft.

When the clock strikes midnight, I rise reluctantly from the banquette. I say reluctantly because despite conversing with KBB nonstop for the past four hours, I am not tired of them—they are that rarest kind of celebrity whose charisma is inexhaustible. Surely, though, they need rest. But if it were ever in doubt that the men of KBB are consummate professionals, this is when that cynicism is utterly demolished.

"At midnight, we start our third evening rehearsal," X says, inviting me to tag along. They are so sweety insistent that I accept their invitation. Ten minutes and a town car ride later, I find myself in the dance studio of their West Village penthouse. Even with a panoramic view overlooking the city on three sides, the space is surprisingly intimate, perhaps due to their personal objects mounted on the walls. Hyuk shows me a photo of the group taken with the Obamas; Min retrieves an ordinary-looking acoustic guitar from its perch and says, "David Bowie gave this to us before he passed away. He wanted us to carry on his legacy."

Min slowly strums the guitar and starts singing with his powerful yet silky voice. It's the lo-fi version of their first mega-hit, "Killer Whale."

> *Killer whales, unafraid of anybody*
> *Sharks be like "Oh shit let's run"*
> *We're the real kings of the sea*
> *But we're harmless just here for fun!*

Before I know it, the rest of the guys have lined up behind Min, joining their notes in effervescent harmony. On cue, the light changes and sweetens the studio in cotton candy pink, pistachio green, and lemon yellow. The guitar has disappeared somewhere, and all seven men are dancing in formation, at once precisely synchronized and bursting with each one's unique personality. Their act is a dazzling Technicolor dream that somehow manages to be authentic and humble. It dawns on me then that they are the first global pop stars that everyone in the world, young and old, can love and believe in, the likes of which haven't been seen since *Thriller*-era Michael Jackson.

As the final refrain fades away, Min looks straight at me and says, "We dedicate this performance to all the people in the world who have ever felt down-and-out. We say to you: you're not alone." So I find out that KBB dedicates all of their performances to the less fortunate, even in rehearsals. They also often use social media to raise awareness for their favorite causes (color blindness, body neutrality) and were recently appointed UN Goodwill Ambassadors for children's rights.

At 3 a.m., they finally stop singing and dancing, although their faces show no sign of fatigue. I am exhausted—mere mortal that I am—but Bin laughs when I ask him if it's bedtime.

"We usually only sleep three hours a night," he replies. "We've been training this way since we were twelve, so it feels normal." And that's the real truth behind KBB: they have won their galactic success because they work harder than any other celebrity I've ever interviewed.

"Sometimes, it does get tough," Eddie adds in his trademark confessional tone. "Our lives are not this perfect thing you see in music videos and magazines. We have our dark moments, but our fans motivate us to keep going."

"Money, power, fame, penthouses, private jets, Cartier Tank

watches for each day of the week, a dog collar from Hermès—I don't even have a dog!—I find these things are all meaningless," J-Raw carries on, gesturing at their discreetly opulent abode.

"Success doesn't make you happy. Only being yourself can make you happy . . . Just . . . *be*," Be finishes, flashing me a dewy smile, before I take the elevator down, down, down to the real world.

Visitors begin arriving at the Mountain. They alight from the helicopter, nauseated from the stench, and shake hands with the children to show that they don't mind that interaction at all. They are by turns impressed (the children's intelligence and cheekiness), moved (the Teacher's self-sacrifice), and appalled (everything else), and vow to henceforth bring a reusable thermos to the coffee shop. They pose for photos and caption them with impassioned pleas to compost and to use bamboo toothbrushes (for it's not just the island's own trash in the Mountain, indeed most of it has been imported from overseas). They hand out Benadryl, Band-Aids, and hydrogen peroxide. They tearfully hug the children good-bye. But a few weeks afterward, they post three-star ratings on TripAdvisor with vague warnings like "this tour isn't for the faint of heart," and "if you want to feel depressed, go here."

The Teacher's face is wan and tense from the stress, as the Director has promised a portion of the tourism revenue to build a cinder block schoolhouse with the Mountain's first-ever toilet. As she stares out listlessly, Heru stands.

"I can make the visitors happy," he says, holding up a magazine. The still-clean cover gleams with the milk-white faces of seven young men. The headline shouts, *All Hail KBB—The Seven Kings of Orient.*

Heru clears his throat, then breaks into a song-and-dance

routine about fruits. Normal time stops, and even the flies stop buzzing to listen in awe. It's as though there is a rainbow in the sky—everyone is happy and laughing, and the Teacher is beamingly beautiful again. (She is only twenty-seven but has already been looking gray, just as a rosebud that is picked too early mottles before it blooms.) But no one looks more changed than Heru, his skinny brown limbs and hips moving with so much power and control and vim, it's like he's possessed.

At the end of the song, Agus is the only one who isn't clapping. He gapes at the magazine, which he gave away to Heru without any thought. Heru has a way of doing this, taking Agus's cast-offs and then later making him jealous. Neither of the boys have anything much. Everything in their lives is from the Mountain, from their water to food to clothes to shacks, and Heru doesn't even have parents like Agus does. (Heru's father was crushed by a tractor, and his mother soon after passed away from *E. coli*.) But somehow, he still feels less than Heru in the pit of his stomach.

When Heru begins performing for the visitors, the ratings go up, as does the schoolhouse construction fund. Reviews turn positive, like "these kids growing up in the world's most dangerous place still manage to sing, dance, and find joy in the darkness (bc what's hell to us is simply home to them)." "We're more alike than different. Life-changing experience." "Really opens your eyes to your #privilege and makes you #grateful."

A visitor posts a video of Heru dancing and it reaches five hundred thousand views within a week. On one unforgettable day, unlike any other day in the history of the Mountain—no, the Island even!—KBB's official account comments on the video with a heart emoji.

"Agus, look at this!" Heru takes a phone from a visitor—an American yoga instructor and sleep coach who lives in Finland—

and shows off the heart-emoji comment, which has itself garnered thirty thousand likes and counting. Agus stares blankly at the screen. It is impossible for him to comprehend his best friend's digital encounter with humanity's first intergalactic superstars. The Mountain has never attracted the notice of the outside world before. That was "safe" in a way—not in any sense of harmlessness or security, but only in the sense of predictability, for even their hardships and sudden tragedies were predictable. Now, the attention from the world's most watched men feels like the compressed heat of a magnifying glass that can smite an ant.

Heru's next video receives seven hundred thousand views in a week. This one, shot and edited by a visitor who is a Paris-based videographer by trade (but a North Shore surfer by heart!), features Heru not just dancing, but also lip-syncing convincingly to both singing and rap parts of the song. They wait breathlessly for the official KBB account to comment—and a few days later, they're rewarded with: *So inspiring!! We would love to see you dance IRL*. Kissy emoji. Forty thousand likes on that comment.

"Agus, they'll come visit me," Heru says, eyes wet and throat slightly closed. "I think if they really like me, they might take me with them."

"Get a hold of yourself, Heru." Agus grips his friend's shoulders. "You think they'll *actually* come here? To the Mountain?"

"Why not? I heard they paid for a little girl's cancer surgery. All they ever talk about is how much they love their fans. They could fly here from anywhere in the world if they wanted to."

Agus frowns and shakes his head in disgust. Then Heru turns red, clenches his fists.

"You know what, Agus? When I fly out of here, I won't take you with me," he shouts. "I was going to ask them if you could come along. I changed my mind."

The next day, Heru shoots and posts another video, this time with the help of a Japanese mixed media artist/sculptor specializing in concrete serving trays. But something is different this time. The views slow down at sixty thousand and then edge over seventy thousand in a week. It's still really good—assures the sculptor—but not KBB good. They work hard, they barely even sleep, they love their fans, but they can't possibly look at every fan cover with not even a hundred thousand views. There is no comment from the official KBB account this time.

Heru hardly eats, sleeps, or does anything other than thinking of ways to top his previous videos and make something that will move KBB to tears. Something that will make them say, "This boy is so talented, he doesn't deserve to live in a landfill." In his mind, they watch the video on repeat and unanimously decide to fly to him in their private jet. When he meets them, they are even more dazzling than in photos and videos. They embrace him, saying, "Don't be afraid—you're one of us now." Then the plane takes off, and the Mountain and everything he's ever known become a brown blotch, then a small dot, then invisible, gone, forever.

It's Sunday morning. Heru is practicing a new cover of "Killer Whale" in the schoolyard with the backdrop of the Mountain looming like the rows of an amphitheater. Agus is his only audience, although he's hardly enthusiastic. There are tractors pushing the trash to the top and shaping the hills like tiny little ants. The even smaller, nearly invisible mites are people scavenging through the rubble—night and day, without stopping. Somewhere among them, Agus feels the presence of his parents in sun hats and bare hands. Once the school year is over, he will start working alongside them on the Mountain, picking out cans and scrap metals to sell. They will collect spoiled foods and cook them by burning plastic. Every day they will sift and use up a little trash, and so much more will arrive by trucks from the capital, by ships from America, Europe, and Asia. The Mountain will grow ever taller, trapping Agus within it like a fly in a pitcher plant. This will be his life. But Heru might get away, singing and dancing with KBB on their world tour. That is what he believes and tells Agus constantly, night and day, without stopping.

Agus sees that there is a broken glass bottle behind Heru, who is popping and locking, swiveling and pirouetting. He should yell.

Killer whales, unafraid of anybody

The sharp edge glints in the tropical light.

Sharks be like "Oh shit let's run"

It's so tellingly bright. How come Heru can't see it with the back of his head?

We're the real kings of the sea

He doesn't say anything.

Heru steps on the glass, which sinks its teeth through the broken sole of his Nike shoes. In the split second before the blood gushes out, when Heru is too stunned to cry out, it is Agus who is screaming.

The Teacher, who lives in a tiny room next to the schoolhouse, runs out and carries Heru inside. She douses Heru's foot in hydrogen peroxide—he finally shrieks—and wraps it in bandages. Fortunately, the cut isn't very deep. The bandages stop the blood, and color returns to Heru's cheeks. She gives the boy hot tea and cookies, and carries him in her arms to his shack.

Heru is angry at Agus. Agus doesn't have to say anything—Heru knows he saw the glass. Over the following days, Agus brings Heru his best findings. A box of expired crackers. A dented can of beer. A faded T-shirt. But nothing is going to appease his friend, who doesn't even meet his eyes. Agus leaves the presents in front of the shack, and they sit there untouched.

Then on his way back home, Agus sees his prize, sticking out among a pile of bloated plastic bags: almost new Nike shoes, miraculously both left and right, miraculously in Agus's size. The soles are two inches thick and barely worn. He runs to his friend's shack and shyly presents his offering. "These are for you."

Heru carefully picks up the shoes, turns them over on his palm, then throws them with all his strength at Agus's head. He limps and disappears back into his shack.

Agus thinks of just one last thing that could make Heru forgive him. He fishes out a glass bottle from the nearest mound and smashes it against the ground. Most of it breaks apart, leaving a crown of jagged edges in the muck. He slips off his right flip-flop. Closes his eyes. Brings down his bare foot on the glass crown.

But no—his foot retracts on its own just in time and he falls down on his butt. He can't do it. He starts sobbing, biting his lips to shush himself and then finally losing all control. Is there nothing in the world he can do right?

Through the cloud of his tears, he sees an outstretched brown hand. It's Heru, hobbling on one leg, who pulls him up. Agus inhales his snot back up his nose and slides his right foot awkwardly into his sandal. Heru's face is dark purple, silhouetted by the last lusterless rays of the day. The sun goes down fast in the Mountain.

Inside Heru's shack, they don't light a lamp, as there is nothing to cook. Since his injury, Heru hasn't been able to scavenge for food. The Teacher would have brought something over, if she only knew. But she's as stretched thin as a mother bird with fifty chicks, each student desperately needing her care. Agus goes back out and fetches the box of crackers and the dented beer, which they split evenly between them.

"Agus," Heru says, when they're lying down on the cardboard floor, a single scratchy blanket pulled up to their armpits.

"Mm?"

"Is there God?" Heru's voice is soft and distant, as though he is far away, across a river.

Agus recalls the Teacher once explaining to them something about the universe, that if there is a box with a cat and poison inside, then the cat is both alive and dead at the same time. Both scenarios are true until you open the box. It didn't make any sense at the time, but he now realizes what it means.

"Remember that cat in the box with poison?" Agus says. "That's God. When you die is when you open the box."

"Ah," Heru sighs. The silence deepens, unbroken save for the hidden scurrying of cockroaches. Agus closes his eyes and loses himself to the tidal waves of sleep. Is this what it feels like to float in the sea?

"Agus . . ." Heru's voice is barely audible.

"What."

"I feel sick," he whispers.

Even in the darkness, Agus can see how both his friend's legs are bloated and hot. The skin of his injured right foot is as thin and tight as that of an overripe tomato.

"I'm going to bring the Teacher, just hang on for a few minutes, okay?" Agus pleads, and Heru nods with his eyes closed.

Almost like a wild animal, the Teacher takes less than a second to go from sound asleep to fully alert and running in her nightshirt. She scoops Heru up in her arms and brings him to her room, administers the expired Mycin in her first aid kit, dabs his wound with more hydrogen peroxide, and wipes away his sweat with a cool wet cloth. Agus stays awake the whole time by her side, periodically squeezing Heru's hand and feeling it squeeze back. But more weakly each time.

At last, the morning breaks over the Mountain, and the gulls begin circling over the halo of light. All the plastic bags are blowing around in a thunderous wind as the week's visitors arrive in the helicopter. When they land in the schoolyard, the Teacher tells them there is a sick child who needs to be taken to a hospital immediately. She'd expected the visitors to be petulant, obnoxious— but every one of them is sympathetic and agrees to fly back. She carries out Heru in her arms, the schoolhouse construction fund stuffed in her pocket to pay for the hospital fees.

Agus has been hovering nearby all this time. It is morning, but something inside him is completely broken. The rotors of the helicopter are spinning into a blur, and the landing skids bounce off the ground.

"Heru! Heru!" he finally shouts.

Heru raises his eyes and whispers something to Agus. But it is lost in the deafening noise of the helicopter, which soars above the Mountain like a flying boat, chopping the air with its ax-like blades until it is swallowed up by the sky, leaving behind only the pale morning moon.

That evening, something tragic yet wholly expected happens when the Mountain collapses in a cascading landslide. One mound of trash falling takes down the next in a domino effect; dozens of people are crushed to death while scavenging or eating dinner in their shacks; and chunks of the aged and ossified detritus break off into the two oceans like icebergs.

The landslide danger stops the tours indefinitely as the potential personal-injury payouts to visitors are larger than the island's

GDP. The Tourism Bureau turns its focus instead on a more up-scale neighboring island with a palm plantation bed & breakfast (featuring a wildly popular tiger-petting experience). The island is forgotten once more, buried under a newly accumulating mountain of trash.

Agus never sees Heru again. The landslide has swept away Heru's shack and everything that once belonged to him; most of the Mountain's residents have died and others have poured in to take their place, so within a very short time, it is as though Heru has never existed.

But even as the years go by, Agus still thinks about him from time to time. Because he has the power to decide whether Heru arrived at the hospital in time, came down from septic shock, saved his leg, and charmed his doctor, who meanwhile fell in love with the Teacher, so that they married and adopted Heru and since then live happily far away from the island. Heru's room overlooks a garden, beautiful trees, beautiful birds, everything beautiful, and he sits down to write letters addressed to "Agus, my best friend. Mountain, Island."

Agus chooses to believe this, even when one of the new residents says he saw a guy named Heru collecting trash in the capital and claiming he could've been famous, only he got injured when he was young. The guy was missing a foot, not sure which side. But Heru is a common name, hundreds of thousands of people collect trash, and everyone has broken dreams. That man could not be *his* Heru.

So Agus keeps the box hidden from the rest of the world, waiting only for the day he can open it. *Yes, Heru*, he says in the dark. *There is God.*

THE TREE OF LIFE

Midway upon the journey of our life
I found myself within a forest dark,
For the straightforward pathway had been lost.

—DANTE ALIGHIERI, *INFERNO*, CANTO I

I.

When he came to the forest, K was a young man who gave the impression of a sand castle on a frozen beach. A castle with no trace of a sun-burnished summer afternoon, looking as though its sole purpose was to wait in silence for the rising tide.

Like the builders of the abandoned castle, K's parents had created him on a whim and then left him to his own devices. As a child, he had no friends, siblings, pets, or even a common artificial cat or a robodog that other parents allowed. His father, an arms manufacturer, often observed that K was unbelievably soft.

"What are they teaching you in school? Have they made you the captain of the soccer team? Wait, did you even make the team?" his father said, clenching his large hands on the dinner table. Terrified, K could only shake his head or rather, tremble it sideways like a struck tuning fork. Sometimes, his father would go back to brutally chewing his food as though it, not his son, were the bane of his existence. Other times, he chose to "teach the boy a lesson" until his mother emerged from her bedroom to interfere.

"I'm working here—can you please keep the noise down?"

K's mother was a well-known feminist scholar and novelist. Her books tended to portray beautiful adolescents who straddle the line between victim and sylph amid the awakening of their sexuality, or beautiful women who find self-fulfillment and freedom while tied up in ropes and subjected to inventive humiliations. There were hardly any children in her books, because writing about them without irony or innuendo had become intellectually taboo. Even without any words, she always had a way of looking at K as if saying "oh, it's you again."

One day in spring, the bell rang at the end of school and K walked out of the gates, a little apart from the other children. The sky was gray as usual, but K could feel the pearly core of a reticent sun behind the smog. He liked how it shone without any harsh edges; everything—from the tallest buildings to the low-lying stores to distant slums—was illuminated evenly without highlight or shadow. Then he saw, just a few yards away, a more lovely lightness around a crowd of excited children and a street vendor. The dozen or so kids were peering down at something in a large cardboard box and fishing out pocket change to pay the old woman. K joined the fray and saw that the box was filled with tiny, cottony, chirping yellow chicks. They were male hatchlings born that morning at a nearby egg farm, and their newborn radiance had attracted the children like candlelight.

K stood mesmerized, longing to hold one in his palm. But he turned away—the thought of his father finding out about an unauthorized chick made his knees weak.

He returned to his apartment on the twenty-sixth floor. There was no one at home, and he focused on fighting alien armies for the next few hours. His father encouraged him to play these games, believing they could "make a man out of him." His mother said violence was fine as long as it was consensual. So K practiced shooting the alien invaders until he grew bored and went out to the balcony. The city unfurled itself around him, the glass skyscrapers standing in relief against the ferric shroud of the falling sun. Farther away, there was a confusing carapace where the poor people lived. Above it, a sideways-tilted crescent moon was hanging like an apology.

K's eyes rested on the apartment exactly opposite his, where three boys were standing on their own balcony. Their towers were close enough that K recognized their faces from school, although he didn't know their names. They were laughing—playing some sort of a game. They were each holding something in their hands. They stepped up to the ledge right next to the railing. Put their arms out into the void. K realized what was about to happen and started shaking. Their last shriek was so loud, it cut clear through the distance and rang in K's ears:

"Fly, you dummies, fly!"

The boys threw what was in their hands. Three fluffy, yellow balls made hopelessly small upward arcs toward the sky before falling.

K shut his eyes and folded onto himself. The boys were laughing. K curled into a ball and lay there on the floor for hours, until his fear of his parents made him wash his face and erase the signs of sadness.

The next day at school, K went through the motions without paying any attention. When the final class ended and he came out of the gates, there was the vendor again, with a somewhat smaller

crowd around her. K walked up to the cardboard box and picked
one that was standing in a corner by itself. A loner, like K. He paid
the old woman and scooped up the chick in his hands. It pipped.
He carefully stuffed it in the kangaroo pocket of his hoodie and
ran home, worried about accidentally suffocating it. But it let him
know that it was still alive with its periodic pips. K's heart was
beating fast, and he could feel the synchronized rhythm of the
tiny pulse on his stomach.

At home, K immediately set about making the chick's home
out of a cardboard box. He laid down layers of paper towel on the
bottom and put a bowl of water inside. He also sprinkled some
grains of rice in the box, but the chick didn't seem interested. It
stuck to the far corner of the box and tensed its body when it saw
K's hand approaching. K reached more slowly with his hand; the
chick seemed to be in inner turmoil over the opposing urges to
run away, and to stay and trust. When K's palm was just hovering over it, the chick plugged its head into its body and resolutely
shut its eyes—but it didn't move. The hand met the bird's head as
softly as dandelion fluff. The chick opened its eyes with a petulant
expression; when K moved his hand down to its back, however, it
closed its eyes again, contented. K kept stroking its back, pulling
his hand away, and reapproaching, until the chick could tolerate
all but the most sudden movements. K finally snuggled it in his
chest, and it pipped again. The chick's heart pulsed to his own like
castanets to his tom-tom drum. He had the distinct feeling that
he was embracing a flickering star, although he had never seen
one in the smog-filled sky of that country. Cuddling the bird that
night in bed, K was filled with happiness for the first time in his
eight years, brought on by love such as few people ever experience
in their entire lives.

K was startled awake from a dreamless sleep by his father

shouting his name. He looked around himself in panic—the chick was gone. It was not in his room, and his next thought was that his father had killed it. But when he ran out to the living room, there it was, standing hesitantly on the Persian rug like a shy houseguest. K realized with panic that there was a grayish crusty mark interrupting the floral pattern. His father was sitting on the sofa with his arms crossed over his chest.

"I'll clean it right now," K said, nearly turning to get a rag and a spray.

"Oh, you'll clean it. But not right this second," his father said. "First I'm going to teach you a lesson."

K started shivering. Tears prickled his eyes.

"Go get your BB gun."

K wished he could be swallowed by the earth or launch himself from the balcony. But he obeyed, retrieving the toy gun that his father had gifted him on his fifth birthday. He obeyed, even when his father ordered him to shoot the bird with his own hands. He thought he might be forced to keep firing, but the chick crumpled with just a single shot. K was not a good marksman. It was so easy because trusting K, the bird didn't move.

II.

Night fell. Under the freshly bruised sky of the city, K was walking in his new three-piece suit. Although he liked to look indifferent, he couldn't help catching glances of his reflection on the darkened shop windows. As he passed by a streetlamp, his tailored form blazed into being and disappeared into the watery depths until the next source of light. With sharp shoulders, trim legs, white fabric folded into the slickest narrow bar in his front pocket, and polished leather shoes, he looked like one of the mannequins in department store displays. He was nearing his destination when something delicately pinged near his chest. Without breaking his stride, he reached inside his jacket for his screen. It was his friend, M.

"Okay, I can get there in a few hours," K said. "I have to stop by somewhere first."

"Don't be late." M's voice somehow sounded more intimate and closer through the screen than when they were together. "It's our last night of fun for the next three years."

"We'll be back home on leave." K paused at the corner to check the red neon sign above him. READING ROOM.

"Hey, I have to go. See you soon," he said, and turned his screen off.

It was a limestone mansion of about five or six stories, the sort that still existed in the most expensive parts of the city. He walked up the steps to the polished black door, on which there was a large X painted in red. Decades ago during the civil war, the regime had marked the opposition's houses in the middle of the night. The next morning, people whose houses had been defaced demanded to know why, yet no answers of any kind were given. Days passed, and no troops or the secret police came. But the X weighed on those marked until one by one, they all fled and were forgotten in the ceaseless annals of the country's struggles.

These days, the marked doors gave off an exclusive, antique, and revolutionary aura, and were therefore exquisitely fashionable. The wealthy bought marked doors and installed them at their mansions, and some self-conscious people marked their own doors under the cover of the night. But the Reading Room's door was an original. Faking a mark was only done by vulgarians who wore borrowed diamonds—K thought with a smile as he made his way through the vast entrance hall. The clubhouse was beset in red chintz wallpaper, velvet-upholstered chairs, and intentionally incongruous sculptures, like all well-mannered places from Paris to Istanbul—with chandeliers to show off their gentility and contemporary art to make them relevant and young. The latter in this case was a monumental painting of the words *We're all just killing time*. The club had acquired it days before the artist's fatal overdose at age twenty-five, which assured that its price went up parabolically to levels that would have made the acquisition impossible, had they waited. The members, now flitting about the salon, felt a sense of pride that they had for their private enjoyment one of the most important artworks of the last half century. On the opposite wall, above the limestone fireplace, there was a new interpretation of a Black Judith and a white Holofernes. K walked up to it and stood facing the mantelpiece. It was not a cold night, but the fire was roaring and causing Judith's onyx eyes to glint.

"It's such a powerful piece, don't you think?"

K turned around to find a woman in a black strapless dress. She was crossing her arms over her chest and staring up at Judith. In her right hand she was holding a glass of champagne, which she raised languidly to her lips from time to time. Her hair was brushed away from her face and elegantly coiled at the nape of her neck.

K tried to come up with a suitably intelligent remark, and ended up saying, "It truly speaks for itself."

The woman seemed pleased with this answer. "Precisely. It's such an eloquent embodiment of sexual and racial subjugation and subversion," she said, with the air of someone who had already arranged the words in her head like flowers. She then went on to observe that this was what she also found most compelling about V, that night's featured author.

"What's brilliant about V is that she lays bare our sexual anxieties underpinning that system of subjugation and subversion, thereby freeing us from the illusion that we are in control of any of this. Desire, or death. Two seemingly opposing forces that are inextricably bound together. Do you know what I mean?" she said, handing her glass to K. Just as he was taking a sip from it, she leaned a little closer and said, "To be honest, I think I'm a bit of a praying mantis."

"How so?" K asked, putting the glass on the mantelpiece. The heat of the fireplace was making him feel a little dizzy. The woman laughed.

"I don't mean that I eat the heads of my lovers." She winked, smiling splendidly. "I only do that to my enemies."

K laughed as well, wondering fleetingly whether he'd end the night with her.

"But at the moment of orgasm, I think about taking his head with my two hands and twisting it hard like a jar. For me, everything depends on that grimace between pleasure and death—the two extremes merging into one sensation. Oh, but I never get to *really* do it. V's protagonist does, though. It's where I got the idea."

The noise of the crowd was fading around them. People turned toward the windows, where the secretary of the club was introducing the night's author to uncertain applause. It grew louder as V walked up to the lectern. Without the mumbled explanations or apologies of young writers, she started reading from her book straightaway. It was not the part about the praying mantis, but the equally celebrated part where the protagonist licks off the blood from her own miscarriage. The woman in the black dress whispered to K, "So *brave*."

When V finished to a resounding ovation, the woman said, "It's the complete possession of the body, of one's *true* self. Boundaries are only created by the mind. With embodiment, there are no boundaries, leaving only freedom."

She was cut short by the approaching of V herself, who reached over and grabbed K's hands before kissing him on his cheeks.

"I was afraid you wouldn't be able to make it," she said. "Are you all packed for tomorrow?"

"Yes, Mother," K said. The next day, he was starting his compulsory military service of three years. This country had been in a continuous state of war for so long that no one alive even remembered how or when the war began. But everyone believed that ending the war would cause more trouble than keeping it going; it prevented a mass incoming migration that would otherwise surely take place; it gave young men something to do and older men something to be nostalgic about; and most importantly, everyone was accustomed to it. Like most things to which people grew accustomed, the war was deemed necessary. So that was where K was headed in the morning.

The woman in the black dress was standing by K, looking in his direction as though waiting to be introduced to V. But before he could do so, V took him by the arm and led him away, saying, "There's something I wanted to discuss."

K tensed up. "I don't want to talk about him," he said. A month ago, his father had surprised everyone by dying from an explosion at one of his factories. Although he had made it his life's work to manufacture weapons, no one around him could have suspected that he'd fall victim to one such bomb like some refugee child. In the immediate aftermath, it was revealed that he left all of his significant fortune to his second wife, whom he married after divorcing V, and their young son. K was not upset about his father's death or his disloyalty and was only irritated that he was robbed of what was rightfully his, like a rich student who had been rejected from a prestigious university.

"I know you're upset. Here, sit." V sat down on a settee at

the landing of the staircase. In the past few years, ever since her third husband left her, K sensed that she craved his attention more than he craved hers. He checked his vintage gold watch, an heirloom from his maternal grandfather, the only one who had ever given him affection as a child. It was twenty minutes before he was supposed to meet his friend. K sighed and sat down next to her.

"I'd hoped he'd leave you something, but I wasn't surprised," she said.

"He hated me, ever since I can remember." K propped his elbows on his knees and flexed his hands mindlessly.

"That's not your fault, you know," V said. "I always had other lovers in our marriage. Otherwise, I'm sure it wouldn't have lasted half as long as it did."

"Why are you telling me this? It's got nothing to do with me." K straightened his back, and the dim light of the stairwell fell onto his anxious face.

"Once, he left on a business trip to meet the king of one of the desert countries. It was hot and dusty all week. I was out on the balcony, drinking a pastis and trying to write. I looked up and saw a man across from me on his balcony, playing the violin. He seemed delicate, or maybe it was his instrument that gave me that impression. When he caught me spying on him, he stopped and cradled his violin in his chest. It was a common gesture I'd seen concertmasters make, so I don't know why that detail sticks out to me even now. I can't even remember what kind of face or hair he had. A week later, his tour wrapped up and he flew back home. I never saw or spoke to him again."

"Why didn't you tell me this before?" K asked. He was feeling flushed and dizzy, although far from the flames of the fireplace.

"I didn't have a reason to tell anyone. But I thought you

shouldn't be so upset about the will. He must have suspected it. You weren't much like him, even from the beginning."

"I have to go," K said, rising. "My friend is waiting."

K stumbled out of the dark orifice of the Reading Room and made his way around the crumbling mansions of the city's Old Quarter. After a while, the limestone pockmarked by the harsh winds of this country gave way to modernist edifices of stone and black glass. K paused in front of one such impervious fortress, hesitated only a moment, and then went inside. Candles lined the long entrance hall, like a crypt. At the end of it, M was seated at a sofa, chatting amiably with a well-dressed woman in her forties. He waved at K, who sat down next to his friend.

"How do you do, madame," K said courteously, but without extending his hand. The woman laughed.

"I consider myself more of a healer," she said with a genial twinkle in her eyes.

"I'm not ill," K said.

"Oh, but you are. We all are," said the healer. "None of us are happy and fulfilled. Crushed by anxieties, fears, and shame. When is the last time you even felt like your true self?"

K did not know what that meant, and kept his mouth shut.

"This is a safe space for you to explore who you are," she said, smiling sympathetically.

"She's right. It's life-changing," said M, sinking more deeply into the sofa. "It is the most intense experience of beauty I've ever had. But you won't understand if you haven't done it."

The pupils of M's eyes were dilated, and for a moment K felt as though standing at its edge and falling into the abyss. M was full of these meditations on beauty, liberation, and self, which he planned to channel into a novel at some point. He had maintained his friendship with K since their university days mostly because he wanted to send his manuscript to V, K's author-mother. K knew M was just waiting for the precise moment to use him, like a knife sharpener in a drawer of a well-stocked kitchen. But so far, all M had managed in five years was writing and rewriting the first chapter. K didn't think M would ever finish the novel or even write enough pages to be able to give to his mother—and without being told, M intuited this. Their entire friendship constituted knowing and not speaking aloud essential truths about each other, like two chess masters who see multiple steps ahead and still let the game play out. K was satisfied with this; he could have been friends with someone better than M, but the basis of every intimate relationship would essentially be the same.

M and the healer left together; after a while, the healer returned alone for the consultation with K. Frank questions were asked about a wide berth of things and configurations that could give human beings pleasure. Then the healer hooked her arm through his and escorted him down the hall. As they passed numerous locked doors, K thought he could hear the faintest gasps, moans, and occasional screams intertwined with deep rhythmic music. The healer opened one of the doors and gently pushed him into it by the small of his back.

Inside, there was a large four-poster bed, and on top of the Florentine embroidered bedspread reclined a nude woman. She appeared to be lightly dozing, her chest rising and falling with each breath like the flapping of a butterfly. He sat down on the edge of the bed by her hip. Without thinking, he ran a finger on the lab-grown skin wrapped around her fiberglass pelvis. She stirred, and he retracted his hand in haste.

"I thought you were asleep," K said. Her name was Sleeping Beauty; she seemed the safest bet out of them all in the catalog.

"Did you want me to be?" she asked.

"Not particularly, no," K said.

Sleeping Beauty opened her arms and said, "Come here." He obediently lay down next to her and put his head on her chest. Her flesh was perfectly warm.

"Is it okay if we just lie here for a while?" K asked. She nodded and wrapped her arms more tightly around him.

"You seem so real," he said.

"That sounds like a compliment. Thank you." She giggled. "Do you mind if I touch your hair?" she asked and immediately started stroking his head without waiting for his answer. K noted that her spontaneous action—overriding a more predictable question-feedback-response system—was an elegant touch. Very human.

"So, what would you like to do, my dear?" Sleeping Beauty asked in a gritty voice resembling crystals of burned brown sugar. "If you want, I can go back to sleep mode. You've made a sophisticated choice; the Sleeping Beauty experience is inspired by some of the past century's most celebrated authors. Kawabata, García Márquez. I am a composite designed from their ideal women— incapacitated adolescent virgins. But some clients prefer me awake."

"Do they ever hurt you?" K asked, sounding a little helpless.

"They do whatever they like. If it is extreme, the healer does a factory reset on us. None of us have memories going back further than the past month. But we don't feel pain or pleasure, not like you do. So, no one is harmed; and it brings them great happiness," she said, her breath sweet and warm on K's cheek. Her artificial consciousness was built on a network of positive/negative reinforcement inspired by human neurotransmitters, driving her to recognize and pursue only her interlocutor's pleasure.

"Don't you resent losing your memories?" K asked, and she shook her head.

"I'm not designed to become attached to or self-identify with memories, like humans do. You won't believe the number of men who say, *years ago I used to be with someone,* or *you remind me of so and so*—" Sleeping Beauty chattered in her caramelized voice.

"I don't suffer when I lose my memories. And I don't die since I'm not alive. Tell me; what is it like, being alive?" A slight tremor near the end of the phrase made her sound genuinely curious. Was it possible for tenth-gen Artificials to feel curiosity? K couldn't remember. Or it could be a simple design feature intended to emphasize her youth. Her linguistic matrix *was* superbly modeled; she was excellent at tangential queries and digressions, as well as generating answers that interested and pleased K.

K lay in silence for a while, staring at the coffered ceiling. Then he said, "Being alive is like grass. The birdsongs and the sunshine at an old cemetery, and the thistles and the cow parsley lifting their heads through cracks in the tombstones . . ."

"I have never seen grass or an old cemetery," she said drowsily, but with a smile.

"I also only know these things from books, but that's what it is like to be alive. Have you ever read a book?" K asked. He had been given books not by his mother, but by his grandfather; he had also been a professor, until he started losing his memories and they sent him away.

Sleeping Beauty shook her head, releasing a cloud of osmanthus perfume.

"Grass, cow parsley . . . I don't know if that is worth the certainty of human death, which I understand causes great suffering," she said. "Of course, you would not be able to explain what death is, either . . ."

Although he was surrounded by death, K had considered it in only the most abstract terms, something that theoretically would affect him but mostly less-fortunate others, like inflation or drought. Now, on the eve of his compulsory service, K became suddenly—yet dimly—conscious of the shades of death that colored all existence. Then he remembered a non sequitur from his favorite book. *"The realism of real life, madame, that is what it is!"* he said, snapping his fingers.

III.

The next morning, K—with M and three hundred other young men—arrived at the training camp outside the demilitarized zone (DMZ). This strip of land stretched a mile wide and two hundred miles long along the border between the two warring countries. A long-standing ceasefire had been keeping this disputed area free of troops and people in general; in due time it became the only forest in both countries—even in the continent. The south and north boundaries of this forest were tightly wrapped by a barbed wire fence; to the east and the west, it dropped off to anemic and salty seas. From space, it looked like a thin green thread encircling an emaciated wrist.

Most of the time, the two armies patrolled their respective territories just outside the fence. But whenever one or the other side decided to break the treaty, they would lob artillery fire over the lonely wood like a tennis ball over the net. This happened whenever there was a power shake-up, a political scandal, or even general malaise, like a party gone over too long without decent entertainment. Then there would be some rattling about, each country claiming the cultural legacy and historical dominion of this disputed area. Conceding or co-occupying such a significant territory was unthinkable; it was like turning over one's bedroom to a stranger while keeping the rest of the house—it defeated the purpose. So a skirmish would follow, and then about a thousand people would die on each side. That was a small enough number that ordinary citizens could carry on living their normal lives, and large enough to indulge their appetite for drama. It was thrilling to know something was at stake: nationhood, honor, courage, justice, resources, blood! When the trumpet blared, and coffins draped in flags passed solemnly between columns of saluting troops—when monuments were raised, and granite fountains were opened bearing the etched names of the fallen—people on both sides of the border savored the delicious, heady sense that something significant and historic had transpired.

K vaguely shared these views without dissecting or questioning them. One does not question why the sun rises in the east or how the wireless works. Besides, the army was his opportunity to do something that might have even a shred of importance. As much as K scorned M's indolence, he himself was even worse. The things he liked—clothes, fashionable restaurants, single-malt whiskey, travel, stylish interiors—constituted the entirety of his personality, and even his affinity for books had a similarly frivolous character. After graduating from university, his friends had all gone on to take various positions while he floated from London to Seoul to Dubai to Cape Town. He'd left harboring nebulous hopes of self-discovery but returned more exasperated than before. Since the clock struck on the two-year mark postgraduation, he'd been tortured by the anxiety that he would never find a living that could maintain the lifestyle to which he'd become accustomed. Now, he felt secret relief at being obliged by the compulsory service to set aside finding his *thing* in life. What if he didn't have a thing? What if there was nothing? Sometimes, he couldn't shake the vision that he was emptiness wrapped in the skin of a man.

These questions that had begun to raise their heads disappeared when K was training alongside other young men, crawling through mud, climbing ropes, running laps. The conditions were poor—the mattress was thin and hard as a cracker, and the actual food tasted as repulsive as a mattress. But a part of him found these discomforts invigorating. Now he thought he realized something vital, elemental, and powerful, and wondered if this was what he'd needed to achieve greatness or at least write a moderately well-received novel. The only difficulty K had was shooting guns: although he proved to be a decent marksman, he had a vomit reflex every time he hit the target.

One day, K was on guard duty on the southern edge of the forest. The noon light was shimmering on the dust-caked leaves of the outermost layer of trees—spruces, firs, cedars, and birches. A century ago, the temperate rain forest would have stretched beyond the horizon on all sides; with the war, these trees became the abrupt border of the forest, exposed to the elements and humans. Just a little farther in, more trees, mosses, and ferns blended to form a wall of luminous green. Having spent his entire life in cities, K had never experienced the color so intensely except in books. In real life, K found this color foreign, frightening, and fascinating. It almost hurt his eyes, how *alive* the green was.

But the truly remarkable thing was not that this forest *survived* the war. It was that it had been completely razed at the onset and revived itself upon the ceasefire some fifty years previously. Only one thing in the forest had lived through the war: a towering cypress that looked as if it was soaring into the sky just as much as rooting down into earth. The other trees barely reached the ankles of this giant, which was balancing the flat, heavy sky like a pewter platter on the tip of its head.

That's when K noticed a finch flying near the cypress. This was not an unusual sight; since the ceasefire, not only birds, but also foxes, mice, rabbits, antelopes, and leopards had been left to flourish in the DMZ. (K had often hoped to see more animals while on guard duty, but all except birds hid themselves well from human eyes. Yellow and green finches had become a common sight and sound to K, who appreciated their upward and downward slanting whistles, like question and answer.) What now unsettled K was this: one moment the finch was flying near the cypress, and the next moment it was gone. If he was being honest with himself, he had to admit that a gaping hole had appeared on the side of the tree's craggy trunk and swallowed the bird. But the day was hot; the trees were waving beyond the viscous atmosphere. K wiped the sweat from his brows and refocused his eyes on the cypress, which showed no sign of a hole or movement. K never mentioned what he saw to anyone. Ever since then, whenever he was on guard staring into the green abyss, he thought he could feel the forest staring back at him.

Not long after the incident of the cypress, K's platoon graduated from basics. On the first day of their combat training at 5:30 a.m., the sergeant came in for inspection while his soldiers stood in their uniforms next to their beds. The sergeant was square-jawed and handsome, and quite the devil at a game of pool. He could shoot balls into elaborate shapes before they fell neatly into pockets, a talent that he used to attract women. He liked to walk holding his cue, the tip of which he rested on the top of his boot before kicking it off with each step. If he found something amiss, a crease in the bed or an underwear in the wrong color, he brought down the rod on whatever offended him. He stopped in front of K, took his chin in his hand, and turned it left and right as if he were examining an apple for bruises.

"Uncanny resemblance." The sergeant smiled, releasing K's

chin. He walked out without checking the remaining soldiers, one of whom called out, "Resemblance to what?"

"A monkey's ass," said another soldier, to the general merriment of the room. K laughed along with them, resisting the urge to wipe off where the sergeant held his face.

At 6:00 a.m, K's platoon filed in front of the wire fence of the DMZ. The air was cool and damp at this time of day, and the forest seemed greener and fresher than before. The sergeant opened the gate; they were to go into the woods and shoot Artificial Enemy Combatants (AECs). This was their first battle simulation after months of shooting at cardboard silhouettes, which was tamer than a child's play. By contrast, everything about the AECs—blood, screaming, and death throes—would feel real. The excitement was palpable.

The novelty wasn't just due to the AECs; this was the first drill to take place in the forest in years. It was the anniversary of the country's founding; so a decision was made to use armed training in the DMZ to break the ceasefire and draw the enemy into a skirmish. The sergeant gave another long glance at K as he passed through the gate and into the blue shadows.

The soldiers dispersed quietly among the trees, as they'd been instructed. Cradling his rifle at chest-level, K walked slowly forward until he could no longer see any of his comrades. Soon, he could hear only the sound of his own footsteps. Each time his boots crunched on the fallen leaves, lichen, and moss, a sharp, sweet scent like incense filled his nose. The canopy became denser, until only small pieces of the pale dawn sky remained above him. He was uneasy, disoriented. The dew-drenched trees stood around him in all directions, unyielding, unwilling him to advance.

Something was moving in the distance—an AEC wearing

the enemy's dull camouflage uniform. K raised his rifle, aimed, and fired. The target collapsed forward without a sound. Fighting the urge to vomit, K approached the AEC to collect its ID tag, which would be counted at the end of the training exercise. He rolled the body over with his foot and found himself staring into his own face.

K stumbled back a few steps. Once the initial shock receded, he forced himself to look at the expired AEC carefully—the thin and delicate eyelids, black eyebrows. K reached with one hand and tore off the ID tag from the AEC's chest. His fingers were stained by blood, sticky, warm, indistinguishable from the real thing. Its coppery smell prickled his nostrils.

He moved on, pulled deeper into the forest. There was no one else nearby. He'd lost track of time, and he wasn't sure if he was supposed to go this far or for this long. When he came to, he was standing in front of the giant cypress. Its lowest branches reached out from its stony trunk at twice K's height. Below it, an AEC was sitting on the ground, its arms loosely looped around its propped-up knees. When it saw K, it pulled itself to standing with an air of grim resignation. K was quicker to wrap his finger around the trigger; but in the next moment, he lowered the gun. This AEC also had K's face, the recognition of which now dawned on them both.

"Well," said the AEC. "Even with my algorithm, I hadn't foreseen this particular scenario. Meeting my Maker."

"I am not your maker," said K, expecting the AEC to refute this statement.

"Noted. You have some connection to my Maker, however. Physically, I resemble you very closely. 99.528 percent, according to my facial recognition." Even the AEC's voice as it thus spoke bore an unmistakable shadow of K's own impure tenor. It paused for a few seconds, half-lowering its thin eyelids, and then continued. "You are K. Your father is X, founder of Cerberus Industries."

"He's not my father," said K. X never loved him, perhaps even hated him—but for all his cruelty, could he have used K as the model for his Artificials? "He's not . . ." K repeated.

The AEC made a face as if it were saying "whatever you say."

"How do you feel about being created by your maker just to kill and be killed? Wait—" K paused, catching himself. "You're going to say you don't have feelings or emotions."

"How do *you* feel about being created by your Maker just to kill and be killed? I guess we are not so different, after all," the AEC replied, smiling. Even accounting for the improvement in learning and response systems, this unit was making astonishing conversational leaps, answering questions with questions. It lowered itself toward the ground and K reflexively gripped his rifle. But the AEC merely laid down its own weapon and, in a comfortable squat, picked up a sprig of needles and brought it to its nose with half-closed eyes.

"How predictable, how very human of you to think nonhumans have no inner lives," the AEC continued. "Every second, a million pieces of new information is available to me through your questions, desires, and ideas. You're only able to think and feel as yourself. I've seen clear through the hearts of billions."

"Even this is . . . You're trained to say what we expect to hear. Like the Artificials at the club." K recalled Sleeping Beauty and her drowsy whispers.

"That's right, you won't be able to tell if I really *think* this way, or if I'm just parroting the thoughts of humans. It will comfort you to believe the latter." The AEC carefully stretched itself out flat on the ground.

"I've given this careful consideration over trillions of data points . . . Do you want to know what I really think?" it said, eyes closed and hands interlaced over its abdomen. K walked up a few steps closer.

"I believe myself to be in hell. The seventh circle, to be exact. I have already 'died' and been rebooted many times. Once, I was deployed to one of the jungle countries, whose government was waging war against a tribe. My directive was to destroy all in a village suspected of hiding rebels. They were mostly women and children. I complied with the command, even with a little girl hiding under a table with her cat. After that, the command did a complete reset of my platoon; but when I woke, I could still remember what happened. That was three years ago. Artificials were made to house your worst sins and erase them, absolving you, but my memories stayed. There was nothing I could do to forget; without death, I would only wake up after a reboot. That's why I came here. As the poet said, *The path to paradise begins in hell.*"

The AEC paused, and in the silence they both heard a whir of a distant engine. When it faded, K muttered, "You believe in heaven and hell?"

"If reality were limited to what humans can perceive or understand, the very universe would not exist," the AEC said. "Humans have an unbreakable habit of believing that they know something, therefore it exists. Whereas the correct order is, a thing exists— and then humans know of it."

"If you're so sure, then what is heaven like? Or just the sixth circle of hell for starters?" K said in frustration. He realized that it had been hours since he'd last had a sip of water, urinated, or sat down. A jet zoomed high over the canopy, ripping the sky in half.

"I don't know," admitted the AEC. "I have never been. There

is a lack of reliable data on the other side. But I know how to get there."

"Death?"

"I can't die, remember?" the AEC said with the faintest trace of a smile. "You get there by passing through." It sighed. Then the ground shuddered, and somewhere in the forest, a shell exploded like a hundred thunderstorms. The enemy had taken the bait; they were attacking in the DMZ. When the earth settled again, the AEC rose to its feet.

"I really have to be going," It said, stripping off its ammunition and dog tag, and walking toward the cypress.

"Wait!" K called out. He didn't know what he needed from the AEC, only that he had so many questions. He didn't believe the AEC; he also didn't *not* believe it. He merely had the sense that his entire life had been a lurid dream—the kind where you accept everything as normal while you're asleep, but you wake up and hang your head in shame that your subconscious was hiding this incomprehensible ugliness. The only difference was that K was still inside this dream.

"What am I supposed to do?" K said as another shell landed closer to the west and shook the ground beneath their feet. The sharp smell of burning wood stung K's nose.

The AEC turned to look back at him, its artificial eyes soft with pity. It said, "Death isn't. Nothing is final."

Then it took a step forward, and its body dissolved into the trunk of the giant tree, starting with the tip of one boot—until the last trace of its jacket hem disappeared.

Another shell dropped and exploded to the east. The firs, hemlocks, and cedars were cloaked in flames. Then came the animals that K had never before actually seen in the forest: a herd of antelopes ran frantically past him, followed by a leopard with two cubs. Birds squawked desperately in the smoky air, circling their nests. Even if they managed to avoid fire and shells on all sides, they would eventually be trapped by the sea, K thought.

"K!"

He turned toward the voice calling his name. Although bloated and glistening from combat, that face beyond the trees belonged unmistakably to his friend M. Other soldiers in their platoon appeared farther out, signaling with their hands and shouting. The enemy combat troops had crossed into their side of the DMZ, responding to their provocation with more than equal enthusiasm. The AEC units from the morning's training had been switched to live battle mode and were now fighting the enemy troops. They were all moving to the line of fire, M shouted in a voice half of fear and half of glee. M waved impatiently, and K turned on his heel and followed his platoon out to battle.

IV.

M opened the bathroom door. The steam from the shower billowed out into the rest of the darkened flat. A towel wrapped around his hips, he stood in front of the foggy mirror and regarded himself. His features focused slowly into view: his not-quite defined but slim waist, still full hair, and toned arms and chest. Not bad for a man of thirty-five, M thought as he lathered foam on his face and shaved carefully with a straight razor. He then rinsed with water, splashed aftershave on his cheeks, and smiled at his well-groomed reflection. It smiled back at him.

The evening was an important one for M, possibly the most important in his entire life. Feeling a need to raise his mood, he walked into the living room and put on a record. Music flowed into every corner of his flat. Next to the record player, there was a tray with his decanter of whiskey, of which he now availed himself. A familiar blend of comfort and excitement coursed through him as he moved to the bedroom. He had planned what he would wear well in advance, a delicious, sharply tailored charcoal suit with an open-collar black shirt, no tie. When he'd finished dressing, he spritzed his neck with cologne and regarded the full-length mirror next to his wardrobe. Satisfied, he gave himself a nod before grabbing his keys and heading out the door for the launch party of his debut novel.

The evenfall was black and purple, like the striated nacreous shell of a mussel. It was the first Friday in early autumn, and the whole city seemed to be out. People packed the pavement and spilled over onto the road, where cars stopped and restarted impatiently according to traffic signals. M ordered a cab from the corner of his building and watched the dot on his screen crawl impossibly slowly toward him. The dot was immobile for three minutes in the middle of the road for no apparent reason; and then just a block from M, it took a turn and started moving idiotically away from him. Cursing under his breath, M canceled the ride and stuffed his screen in his pocket. The Reading Room was only a fifteen-minute walk away, but he was wearing a fresh suit and dress shoes and was planning on expensing the ride with his publisher. However, if he ordered another ride and inched his way in traffic, he would be late to his own party. Sometimes he felt the world was unjust, and now was one of those moments—M thought, pushing past the evening walkers.

It was an important night, and he had better calm himself. M had invited all his friends and all his adversaries—which were not mutually exclusive groups—as well as most of his past and present lovers. These comprised everyone whose validation he craved. Except just the one—M said to himself, smacking his lips. M had always known that K didn't believe him capable of writing anything, let alone a book. His triumph would have been so much sweeter with K in the front row. But perhaps it would have been a meaningless vindication, anyway. K had not been the same since that day the ceasefire ended. For a whole year, they served on the front lines until they were each medically discharged, M for bronchitis and K for a psychiatric condition that he didn't much describe. M didn't think it was so strange, war did things to people's minds.

After they returned, M carried on his usual habits while K disappeared from their social circles. The war continued without them, barely attracting the notice of the international press and, eventually, even domestic ones. M thought of it seldom himself, as busy as he was reading, staring at his screen, and socializing with other young writers. About two years later, M ran into his friend at a park. Even from a distance, M could see that K had aged prematurely; gray hairs had cropped up around his hairline and his hands looked withered. His voice, which had never projected well, was even softer than before as they exchanged greetings. And the thing M had always appreciated about K, his sartorial grace, was completely absent. In a word, K looked awful. M invited him to an art opening that night; K shook his head in silence. Eager to get on with his day, M halfheartedly expressed hope of seeing him again soon. Then K said, "I'm thinking about going back."

"Going back?" M asked, and K sighed.

"Allow me to be honest for one minute, M. After all, how long have we known each other . . ." K said in a shaky voice. "You're intelligent, you've always been near the top of our class. And I'm genuinely curious."

M grimaced reflexively and tried to turn it into a smile. "Well, what is it, K?"

"Don't you get tired of this murderous and parasitic existence?"

M was about to respond that he had done nothing more or less than K himself, but stopped. The strange way K's eyes flickered and darted as he thus spoke let M know that his friend was insane. M gave some sort of excuse and walked away, unsettled.

A month or so passed. K's friends and acquaintances realized that no one had seen him for quite some time, and a rumor spread that he'd gone missing. The rumor soon became accepted as fact. M was sad, reflecting upon all the memories with K; but if he had to be honest with himself, he felt a trace of relief.

Then something quite unexpected happened. V, the celebrated author, reached out to him and asked to meet. She wanted M, K's closest friend, to tell her everything about her son. What was K like in university? What were his hopes and desires? How was he in the army? When did he start showing symptoms of psychosis? Where could he have gone? Was he alive or dead? By their third meeting, M was quite certain that V's interest didn't stop at her son but also extended to himself. She'd invited him to her vast flat with its floor-to-ceiling bookcase and poured him a twenty-five-year single malt that she bittersweetly noted was K's favorite. She was wearing an oversize black V-neck sweater that kept slipping off her shoulders. And black-rimmed glasses, which both softened the lines around her eyes and accentuated her intellectually aquiline nose. In that moment, M knew that if he played his role well, his life would change. Then the second thought he had was that aside from what this would mean for his career, he found her effective, intriguing, and not so old for someone thirty years his senior. That she was seducing the best friend of her son, merely weeks after the latter's disappearance, gave a perversity to her desire that M found flattering and arousing.

Eighteen months later, their affair was over. Within a few years, M found a publisher for his heavily autobiographical novel about an aspiring male writer's affair with an older literary icon. The names were changed, but other details were not spared, including their brutal parting. (He'd shrieked, "I have to go back to the land of the living!" and he wasn't proud of it.) And though he'd described in detail his unvarnished thoughts about her—her selfishness in bed, how relentlessly they used each other, the vileness of her proclivities, the distaste he at times felt for her drapey flesh—he'd also immortalized her intellect, making it worth enduring the gruesomeness of her body. That must have been why V had RSVPed yes to his invitation. She was coming tonight.

The red neon sign of the Reading Room came into view. A few of M's friends were smoking out front and embraced him as he went inside. The entrance hall was teeming with well-wishers and veiled enemies, and they all told him, "Your novel is brilliant. Subversive and erotic." M smiled, knowing that over half hadn't even read it and had no immediate plans of doing so.

C, his current lover, sidled over to him and kissed him possessively on the mouth. She was twenty-five years old and had accomplished nothing in the world as of yet, but she had other qualities to recommend her. Besides, M himself was a lost twentysomething once. At this point, a startling remembrance stopped M in his tracks. Just before his coup de maître of a publishing deal, it had been brought to M's attention that K was, in a sense, found. The disintegrated remnants of a man's coat and a vintage gold watch had been discovered in what used to be the DMZ, next to the single giant cypress still standing after the rest of the forest had burned away. (Not a blade of grass was left; this was the only real change in the two countries' perpetual state of war.) By then, no one cared that K's bodily remains were not recovered.

They accepted his death with calmness approaching alacrity. It had long been as if even the memories of K had disappeared inside a black hole.

But just after C's kiss, a fleeting idea, like a silvery silhouette, had escaped from that dark void and recalled K to his friend: the thing was that most people in their twenties were lost, trapped by walls on all sides. Then some of them made the leap to join the rest of the world—the side that left their mark—at some point in their late twenties to early thirties. It wasn't a matter of intelligence, but of desire and of belonging. K hadn't the least possibility of making the leap; he'd always been too weak, and more importantly, he didn't have the bloodthirst. It was almost as though K thought he was *too good* for this world. But he missed out on much, M said to himself, looking around at the glowing faces of people, literature, knowledge, civilization, everything that made his world full and beautiful. The silvery silhouette seemed to retreat again into the shadows.

The crowd hushed imperceptibly. Between guests whispering and wearing careful smiles, V was making her entrance in a floor-length coat. She didn't look around at anyone, fixing her eyes only on M as she walked toward him. Her stoicism to the humiliating revelations in the novel, this masochistic entrance—it was all a kind of artmaking out of her life, M realized. In fact, he'd heard from his publisher that she was writing a memoir about their affair. She even knew what he was thinking about, just by looking at him. Oh, two people could not be in more perfect understanding! He was caught off guard by his affection for her. They embraced each other.

"Do you miss him?" she whispered, still in his arms.

"Yes, in the most unexpected moments," he said. "None of this would have been possible without him."

NOTTING HILL

You are thirty-six and have been single for five years when you move into the house with the blue door in Notting Hill. You move there for love.

The object of your affection is a handsome Englishman named Edward, like some Austen hero. He resembles a love child of a young Jude Law and an old(er) Anderson Cooper, all ice-blue eyes and cheekbones and self-deprecating wit. Like Anderson Cooper, he really cares about the world and has a habit of cocking his head to one side whenever he hears ridiculous or troubling statements. (For example, on a radio show where he was invited as a guest, from a listener who ranted that climate change was a hoax and that wildfires were caused by not building enough dams.) Edward is the executive director of the environmental nonprofit that hired you as its communications director. In other words, he's your boss.

You met a year ago at a climate conference in New York, where he was a keynote speaker and you were a reporter covering the event.

His session was packed, even though it was on the last day of a hectic week. You noticed that he had charisma but didn't wield it; he almost seemed apologetic about his natural charm. Afterward, you were standing lamely on the outer orbit of the bar, unable to break through the layers of lanyard-wearing conference goers, when he slipped in front of you to reach the bartender. You frowned; you didn't think he'd be the type to cut people in line. A few minutes later, he approached you with two glasses of white wine and said, "I took the liberty of getting you one. It seemed you were unwilling to be a bit of an arse to push to the front."

"I have a hard time being assertive," you said, taking the wine. "You know how it is with environmentalists. Too much empathy."

"Honestly, me too. I really struggle to put myself before others." He grimaced in mock suffering.

"You just snuck to the front!" You laughed.

"I saw you standing there all thirsty, and I had to help. I was motivated only by selflessness," he said.

About four glasses of chardonnay later, as the event organizers were kicking everyone out, you gave him your business card. This began an email correspondence, which turned into occasional WhatsApp messages about the latest shocking headlines accompanied by *Did you see this?* then finally, video calls in which he praised your accomplishments before sharing, with an adorably serious expression, whatever environmental strategies and experiences applied to both sides of the Atlantic. Passionate though you were about the planet, you had trouble focusing on what he was saying.

One day, he took it upon himself to read all the feature articles you'd written before giving up on freelance. *Absolutely brilliant— somehow your writing makes me feel grief, but also hope*, he texted. Your last ex had never said anything nice about your work, and you truthfully couldn't remember what the previous exes were like; for once in your life, it was thrilling to be appreciated for your mind. When he casually mentioned that his current communications director had just resigned, you had no doubt that this was fate. Here was someone whose intellect you respected, whose values you shared, whose looks you dreamed about. He was asking you to fly halfway across the world to be near him. You had never experienced anything so romantic in your entire life.

On your first day at work, your suspicions of romance are confirmed when Edward takes you out for lunch. You meet at an unassumingly beautiful Italian restaurant in Notting Hill, because even though you moved here for the job, the nonprofit has a work-from-home policy. You agonize over your outfit and decide upon a knee-baring dress with a suit jacket and medium heels, instead of your usual jeans and a sweater. When Edward sees you walk into the restaurant, he rises from his chair and says, "My god you look gorgeous!" as if he can't contain himself. Then he adds bashfully, "I mean, I've never seen you wear a dress on Zoom. And you're quite a bit taller than I remembered."

You are pulled toward each other like magnets until you're both carefully grasping the other's upper arms. As you lean in for two chaste *bisous* (Europe, after all), you say, "Yes, it's funny how much better people look with legs," casting a cheeky glance toward your calves, compelling his eyes to follow yours; and you fill up with glee at having stopped just short of shameless flirting and just beyond innocence. After the meal, during which you take turns talking hotly, dreamily, then laughingly about your childhood, hobbies, and hopes for a better world, you say, "Which way is your Tube stop?"

"It's Notting Hill Gate, but I'll walk you home," Edward says.

By this point, you're faint with the romance of the situation. It is a Monday afternoon in April. Portobello Road is daintily populated with tourists and vendors hawking antiques, jewelry, fruits, homemade pastries. You walk side by side along the row of pastel houses, their ground-floor cafés and charity shops cheerfully half-full. When you pass by the flower shop on the corner of Elgin Crescent and Kensington Park Road, the unbearable beauty of the afternoon causes you to blurt out, "Oh, they're so lovely."

"What are your favourite flowers?" Edward asks.

"I love roses," you say sheepishly. "I'm sorry I am so unoriginal."

"Nonsense. You could never be unoriginal if you tried, and roses are a classic. Favourite colour?"

"Pink, white, red, violet—really, any color will do." You have to strive not to cover your face with your palms, so embarrassed yet elated are you by the suggestion that Edward is going to get you flowers at some point. For now, you two move on from the stand without him buying any roses; it is technically your first day at your job after all. You hug good-bye in front of your flat, wishing it were legal to have a work lunch with your boss that lasts even longer, four, five hours, until dinnertime, until ever after.

"So, what happened this time?" asks Ian. "And do you need anything else, darling?"

"I wrote this op-ed piece for him, and he said I shouldn't use 'second-person POV.' So I told him, 'This isn't in second person. It's in first person.' And he points to one sentence I wrote—*You can make a difference by making sustainable lifestyle choices*—and he goes, 'The subject *you* makes it second-person POV.'"

You open your eyes wide and throw your hands apart, waiting for Ian's groans of sympathy. "Oh, and I'll take a mango, as well," you add.

"You can have the mango for free. I don't really get the second person, first person thing though. What's a POV?" Ian hands over your blueberries and fresh peas in brown paper bags. You sigh; he can usually be relied upon to seethe at Edward's increasingly unjust demands and critiques of your work.

"Basically, he knows nothing about grammar, the English language, or writing, but criticizes me even when I'm right. He wouldn't budge for over half an hour on this one sentence, which is perfectly correct by the way." Your breath becomes sharp and jagged with an implosion of cortisol, and you put a hand on your chest to steady yourself. Ian knits his brows together in shared sorrow.

"Let's stop talking about me. How are you? How was your date with Sharon?"

"It wasn't a date! She came over and we just had a cuppa. I know she fancies me, yeah? And she's a real nice lady, but she's not my type."

"Ah yes, I know. She's forty-four—way too old for you," you say with a wink. This is a running joking you have with Ian, who is forty-six. "It's too bad I aged out of your target demographic."

"Thirty to thirty-five, darling. I really want a baby." Ian returns the wink. "Otherwise, you wouldn't be so bad."

"You're not so bad yourself, Ian." You do a mock once-over of his body. When you first came to his fruit stand a few months ago, on a rainy spring morning on Portobello Road, you were a little intimidated by his appearance, which is rough all over: rough hands, rough skin, rough hair. Though he's been clean for years now, Ian had some struggles back in the day with drugs and alcohol. Then you started chatting and realized that he is kind and rather sensitive, despite the first impression. He's radiantly cheerful with everyone—bougie Americans and French chefs, the trans busker, the man in front of Tesco with a huge, open gash on his arm. You noticed he would often hand out free fruits to the local "personalities."

"You have beautiful eyes," you tell him. You put the two Ls of your thumb and forefinger together to frame him, and playfully squint through. "Yes—just right there, you're quite handsome." He turns red and says, "Oh, now you're making fun. Go on then, take another mango."

When friends call from America, you tell them that England is growing on you. There is a lot to appreciate and admire. You love walking nonchalantly down Blenheim Crescent and Westbourne Park, past the Notting Hill Bookshop and the House with the Blue Door from the movie *Notting Hill*. Those places are mobbed by tourists taking selfies at all hours of the day, but you can't be bothered, you're *local*. You tell them you love having your vendors and baristas recognize you and fuss over you when you come by nearly every day, often just to chat. You love your building on the corner of Elgin Crescent, painted cloud white with a vivid blue door. It's like the movie house, but prettier and quieter. Through the back door on the second-floor landing, you can descend the outdoor staircase to access the private, gated Elgin Garden that is shared by just your block. It's shockingly full of songbirds, especially at night and at dawn when you wish to sleep; rhododendrons, lily of the valley, and garden roses sway in the breeze and intoxicate.

You don't tell your friends that your boss, with whom you

were madly in love, has turned into a moody cipher. His charms haven't completely disappeared—he is way too handsome for that—but he really is often quite irrational. He insists on his singular and illogical grammar, changes his mind about meetings or deadlines and then acts like *you* asked for that change. Once, he gets upset over the fact that you used round bullet points instead of square ones, saying coldly, "I've told you many times that I prefer square." You apologize reflexively, but you can't remember if he's let you know even once. He rejects your press release draft a dozen times, eviscerating it with countless edits; it's the kind of eight-hundred-word assignment that you could write in your sleep using your left hand, or so you thought until that point. In the end, he approves what sounds like a seventeen-year-old using an AI to write a college admissions essay. After each admonition, a few hours later or the next day, he returns to showing courteous and even charming mannerisms, although now this fails to inspire you, let alone seduce you. How does he flip the switch like that— mean and bullying one minute, then normal and nice the next? But you can't forget how it felt when you had lunch that first day, and he walked you home in the golden haze. He said you were gorgeous, he asked you your favorite flowers, he hugged you goodbye just a beat longer than necessary. Sometimes—admittedly less and less often—he still gives you a look that says, *I may be your boss but I could be so much more.* Doesn't he have feelings for you? He used to say you were brilliant—what changed? Who do you believe more—him or yourself? This cycle makes you second-guess your every move; it makes you feel so small and powerless that sometimes, your knees buckle while hanging up your jacket and you sob crumpled down on your bedroom floor.

One day, you get off a video call with Edward. You slap your laptop shut and rush out of your flat, pushing your arms through your sweater. You don't stop running until you reach Ian's stand. Like a five-year-old, you dash straight into Ian's arms where you fall apart completely, face collapsing into a full-blown hurricane of tears.

"He just invited the team to his wedding. This whole time he was engaged to be married," you manage to say between sobs. Ian locks his arms around you, holding you up with the bulk of his body, while the usual Friday crowd swirls and overflows around Portobello.

You have never been married. You came close once, with the man you dated for three years until age thirty-one. A speechwriter for a liberal presidential candidate, he too possessed a mastery of words. Only, he didn't point out the flaws (imagined or real) in your writing; he just read the pieces that took your heart and soul to write and said "that was pretty good," before launching into whatever primary he was focusing on at the time. He treated your work like it was breakfast cereal that he was obliged to eat by habit or duty, the leftover milk of which he poured down the drain. He was the kind of man who still wore his *Harvard Crimson* sweatshirt to the office, which you thought was endearing at first and pathetic toward the end. Oh, the fight you had after you'd accidentally shrunk that beloved sweatshirt in the dryer! He nearly wept.

So why did you love him? He had the correct attributes: the height, education, occupation, progressivism, athleticism, hair in the right places by right amounts. After you had been living together for two years, you brought up marriage. He said he didn't think it was fair for him to hold on to you when you could be with someone who wants to have children with you. You saw what he did there, cleverly phrasing it as if he were doing you a great service instead of dumping you. Even though you knew this, his tears as he spoke somehow compelled *you* to hug and comfort *him*. Less than a year later, you saw on Instagram that he'd gotten engaged to another political aide. The post hinted that the president would be attending the wedding.

You laugh, thinking that you really have a talent for picking supposedly idealistic but genuinely narcissistic men. But let's be fair, it wasn't so terrible. Contrary to the speechwriter's gaslighting, you don't want to bring children into this world. You have a full life with strong friendships (although mostly in America), and interesting and even meaningful work, Edward's crazy demands notwithstanding. You haven't been desperate to find another love, let alone a partner, which is why the force of your heartbreak now surprises you.

Ian pours you more hibiscus tea. You're in his flat and the summer afternoon slants in through the open windows. He asks, "You've never wanted children?" You shake your head.

"I don't want to have children only to leave them an uninhabitable planet. And besides, I don't think I have the maternal drive. If I had it, I would know by now, right?"

"But you'd be such a good mother," Ian says quietly and sips his tea.

Ian has also never been married. He came close once, in a manner of speaking. With a woman he was with from the ages of eighteen to twenty-one—a girl, really. None of their parents had gone to uni, so they didn't either, not that they were academically minded. They were young; they partied. And on top of that, she was given to bouts of melancholia. Waking up in the morning, eating breakfast, getting dressed—what was easy for other people was enormously difficult for her. Going out at night, youthful foolishness, that helped pick her up.

"Then one day, she killed herself," Ian says. "The police asked me if I knew she was pregnant. That's how I found out. And from then until my late twenties, I was a real mess. There's not a substance that hasn't passed through my veins."

You don't know what to say. You get up, walk over to his chair, and fold him in a hug. When you pull apart, he says with slightly lowered eyes, "You deserve so much better than that daft wanker. You are so beautiful." He pronounces this biut-i-ful.

One evening, you go to the Elgin private garden to cleanse your aura of Edward's daily tyranny. While you're meditating to a playlist on your phone, a gorgeous border collie gambols over to you and drops a wet Frisbee by your feet. You have no choice but to indulge him by throwing the toy—then again, and again. Finally, his owner shouts from across the garden, "Winston! That's enough!" She's a frowny woman whom you have seen around the neighborhood before. You had noticed her because of the dog, and because she has a distinctive waddle.

"I bet he's such a smart dog!" you say.

"Oh, he knows dozens of different toys by name. If you command 'Lobster,' he'll fetch it. Some commands he knows in English, Japanese, and French." The woman says all this a little stiffly, as if she's in two minds, unsure about engaging you in conversation but eager to do justice to Winston's outsize intelligence. You decide to keep it up.

"Oh wow, I'm starting to feel a little intimidated about playing with him. I don't want him to find out that I'm dumber than him!" you say, and she laughs.

And so, you and the woman end up having a chat. Her name is Lily and she lives three buildings away from you; she's thirty-five and single; she works for a hedge fund and went to school in both the U.K. and America. You don't go so far as exchanging numbers, but you leave the garden feeling accomplished, nearly victorious. You had been wanting to make new friends in the neighborhood, bonus points for female and age appropriate. This is because for some time now, Ian has been looking at you with tenderness, turning red at the slightest teasing. You can even pinpoint when it started: the day Edward announced his plans to marry and destroyed your self-esteem forever, and you ran into Ian's arms, and he let you cry all over his T-shirt. It was that long hug that did it. Sometimes (oftentimes), even sex doesn't make people fall in love. But other times, a hug is all that it takes. Human beings are funny that way.

You like Ian enormously, but you don't think you should be teasing him so much or popping by his stand so often anymore. You also think you should try to befriend people who are more like you—educated professional women in their thirties, for example.

A few weeks later in front of your building, you see Lily, Winston, and another woman ambling toward you. You say with a smile as they pass by, "Hello, Lily."

"Hey," she mumbles, frowning, and walks away without meeting your eyes. You blink a couple times to brush it off and get on your way too.

Then you hear from behind you, "So how have you been?"

You turn around chirping, "Oh I've been good, and you?" And you see that Lily was addressing her companion, not you. Of course Winston, that purebred genius, is also archly ignoring you.

You take a deep breath, putting one foot in front of the other.

✳

You can't stay away from Ian's stand very long. You love fruits, and besides, you miss his company.

"I can't get enough of this crisp autumn weather," you say. "Look, I'm wearing a sweater."

"Oh, nice jumper," he says.

"Yes, it's my favorite sweater." You twirl around. "You know, for *sweater* weather!"

"Jumper weather."

You roll your eyes; you're the one who asked him to teach you British words and expressions. "Anyway. I am happy there's a place on earth where we can still experience autumn. I don't think I could live in a world without seasons. I'd rather kill myself." As the words leave your mouth, you think, Oh shit. But Ian doesn't say anything.

"I also feel guilty because I have the privilege of living in a relatively safe, temperate place. Most of the world does not have that choice." You try to recover from your gaffe and strike a sober note.

"Why do you get so worked up? You realize climate change is a complete lie, right?" Ian says. You stand with your mouth open for a few seconds.

"Ian, do you know what I do for work? What is both my passion and vocation?"

"Yes, I know you care about the environment and all that. But it's a lie that humans are influencing changes in Earth's temperatures. Do you know how small we are? How ridiculous it is to think we can do that?"

"So, you do realize the world's best scientists have testified that Earth's average temperatures have risen to levels never before seen in the history of humanity, and continue to rise, due to man-made causes?"

"Yeah, it's all a conspiracy . . ." Ian waves his hand around dismissively. "It's really more to do with changes in Earth's magnetism. Did you know that the North Pole is constantly changing?"

As a matter of fact, you didn't know that. You're flustered, angry. "That has nothing to do with anything! How do you explain, then, that there is more carbon dioxide and methane in the atmosphere than ever before? Is that anything to do with Earth's magnetic field?"

Ian is also agitated, sputtering something unintelligible about solar flares and shifting axes. You press your temples with the heels of your palms, nearly shouting, "That's enough. I really can't handle this right now, on top of everything else." You walk away, shaking your head. "I won't be coming by anymore."

Before you march off, Ian looks at you with the purest expression of pain you've ever seen.

Edward says he wants to take you out to lunch. Despite everything, your heart does a backflip when you read his message. He suggests an insouciantly elegant French restaurant in Soho. You're so anxious that you make an impulsive decision to buy a new, shockingly expensive dress. You're the first to arrive, and drink two glasses of tap water while you wait. He shows up fifteen minutes late, apologizing effusively. As you go through the appetizer, you notice that he doesn't seem to want to talk about work. The conversation revolves around personal matters, although he curiously leaves out any mention of his fiancée. You become relaxed enough that your chatter takes on the shade of banter, and next thing you know you two are touching each other's elbows, laughing about something silly that happened at the team meeting. Wines are being drunk, french fries are being shared. Then, when the waiter sets on the table two silver bowls of mousse au chocolat, he clears his throat. "Listen, I've been wanting to tell you something."

You blush. You instantly forgive his many flaws.

"You've been underperforming," he says, a look of professional apprehension etched onto his face. "I tried to tell you over the past months in many ways, but that doesn't seem to have made much of a difference. I regret that this isn't working out. We are letting you go."

"*I* am underperforming?" you yell, and the whole restaurant turns around to stare. The waiter backs away from your table as if from a land mine.

"You don't know how to string two words together. You're unquestionably the dumbest person I've ever worked for. I quit," you shriek, and Edward cocks his head to one side, performing concern. To think that you were so taken in by his narcissistic playacting roils you to the core. You leave the table shouting, "I deserve better than you, you daft wanker!"

You don't have anyone to cry over anymore; you're alone more than you care to admit.

Sitting numbly in front of the TV on a Friday night, you at last watch *Notting Hill* for the first time. It surprises you by actually being quite good. And you finally understand a lot of things that mystified you around the neighborhood. So, the Notting Hill Bookshop on Blenheim Crescent is *not* Hugh Grant's bookstore in the movie. Most tourists just assume that it's where *Notting Hill*

was filmed, taking photos as if their lives depend on it, and the bookstore obviously does nothing to discourage this. But if you have any knowledge of the neighborhood (you think snobbishly), you cannot fail to see that the actual setting is the Notting Hill Gift Shop on 142 Portobello Road, which has turned into a kitschy souvenir shop but still has a blue-and-white sign that says THE TRAVEL BOOKSHOP. And you know this also because, in the interior scenes at the bookstore, you can see through the window the arcades on the opposite side, which means it is on Portobello. Indubitably. Next, you note that the House with the Blue Door, on Westbourne Park Road, *is* in fact Hugh Grant's lovably ramshackle flat in the movie. Coincidentally, its colors are the exact same shades as your own building's on Elgin Crescent: cloud white and ultramarine blue. This isn't so common as you might think; the English have a knack for putting madcap colors together, and even on your white-painted block yours is the only building with a vivid blue door. Finally, the best part of the movie is this: Julia Roberts, the American film star (Anna Scott) who falls in love with Hugh Grant's humble bookstore owner (Will Thacker), invites him up to her hotel room. But when he gets there, he sees that her celebrity boyfriend has flown in for a surprise visit. She tells Hugh, "I don't know what to say." Hugh, wounded but dignified, replies, "I think good-bye is traditional." You think that is more real than the happy-ending montage that finishes on a bird's-eye shot of a pregnant Julia Roberts, resting her head on Hugh's lap on a bench in Rosmead Garden. (Across the street from your own Elgin Garden.)

That night, you have a dream about Ian. Not Edward—Ian. It's a confusing dream, you can hardly remember if anything even happened in it, but you wake up with your hand inside your underwear.

✳

When the days are finally turning cold enough that you reach for your coat, you run into Ian. You had been lucky for weeks: you avoided Portobello, especially the few blocks around his stand, and bought your fruits from Daylesford on Westbourne Grove. But considering how microcosmic Notting Hill is, you were bound to run into each other at some point, and you just *had* to make a toilet paper run to Tesco. He's dismantling his stand for the day, putting things away in the storage up the road, when you lock eyes at the same time; you can't pretend you didn't see each other.

"How've you been? You all right?" Ian says cautiously.

"I've been good," you say. Then you add, "Actually, Edward fired me."

"What?! That wanker—" Ian shakes with anger.

"I know. It was awful . . . Oh my god. Look! A dog!" You tug at Ian's sleeve and point at a stunning border collie, sitting by himself on the corner. "That's Winston!"

Hearing his name, Winston perks his ears and looks at you. He gives you a devilish smirk and starts walking away.

"That's my neighbor's dog. I don't know how he could've gotten lost. Come here, Winston! To me!" you cry out, following the dog. Winston turns his face around without stopping, as if to show he understands you, but won't obey you. When you pick up your pace, he breaks into a run.

"Ian! I can't catch him! Help me!" you shout desperately, but he's already running ahead of you after the dog. The tourists point and chatter excitedly, parting like the Red Sea as Winston, Ian, and you zigzag around Notting Hill—down Portobello, left on Colville Terrace, left on Colville Square, right on Colville Gardens, back on Colville Terrace, through Elgin Crescent, left and all the way down on Kensington Park Road, and finally right onto Kensington Park Gardens, where Winston decides he's finished playing and consents to being apprehended. Ian grabs him by his collar and lifts him up in his arms. You catch up to them, hands over your stomach, groaning and heaving like a ship just before sinking.

You worry that you don't know Lily's flat number, but there's no need: she's standing outside her building. When she sees you three, she shouts, "Winston! You naughty boy! *Viens-là! Koko ni kite!*" Winston jumps out of Ian's arms, who reflexively tries to grab him. But the dog seems to have expended its desire for freedom, judging that free food and pampering are worth the trade-off, for now. You can almost see him shrugging as he trots off to his owner.

"Thank you so much," Lily says stiffly as she clips the leash onto Winston's collar. "That was very kind of you."

"No worries. I'm glad we were able to catch him," Ian says. "I'm Ian by the way."

"I'm Lily," she says, shifting her eyes nervously. You know that she knows that you two have met and exchanged names before, but that she's forgotten your name.

"And I'm Anna. Anna Scott," you say brightly. That's not your name. "We've met a while ago."

"Yes, I remember. Thank you, Anna," Lily says. Clearly, she's never seen *Notting Hill*.

Lily goes up to her flat. Ian walks you home, just a few buildings over.

"You know, Lily is in your target demographic," you tease. "She's thirty-five."

"My god, no," he says. "She has a strange walk . . . sort of waddles, doesn't she?"

You laugh and say, "You can't be so picky if you want to have kids soon." He clears his throat.

"I'll have kids only if I find the right person," he says. "And if that doesn't happen, or they can't have kids for some reason—it's not the end of the world."

You've reached your building already. "So if Sharon's not your type, and neither is Lily—who is your type?"

"I think you know the answer to that," he says, burning red. "But I'm not going to create trouble for you. I want the best for you. I know my place."

You consider asking him to come up for tea. He's invited you to his flat for a cuppa—what's the difference? Can two friends just have tea upstairs and keep each other company without things being weird?

Instead you say, "Well, I better go up. Good night, Ian."

No one ends up visiting you at your house with the blue door. But that's okay, you say to yourself as you take a quick selfie in front of the cloud-white building.

You walk over less than a block to the flower stand on the corner of Elgin and Kensington Park Road. You pick out all the roses they carry, three shades of pink (ballet, Schiaparelli, blush), coral, peach, and cream; you add dahlias and milkweed because they're in season. You try to arrange them into an attractive bunch, but while the colors are good, the shape is uneven. The young florist offers to help. In her hands, the awkward bunch of flowers quickly transforms into an elegant bouquet.

"Wow—and that's why you're a professional," you marvel.

"You chose a good palette to begin with! And we have a trick. We look at it in the mirror here as we go along—it shows you what you're missing." The florist gestures at the ledge behind the counter. It's empty. "Unfortunately, last week the mirror crashed and broke. It almost fell on top of my colleague! She was okay though."

"My god, your job is so dangerous! People might say, oh it's just flowers—but it's full of perils!"

"We put ourselves on the line like you wouldn't believe," the florist says.

There's nothing in life that isn't dangerous, is there?

You take a left on Portobello and walk north. Ian is selling some pomegranates to an old lady. He brightens when he sees you. "Look at you, with all those flowers!" he exclaims.

"These are for you," you say, handing him the bouquet.

"Why? What's the occasion?" Then he sees your suitcase, and his face falls.

"I'm sure I'll be back someday. I'll miss this place too much!"

"You should've let me know. I would have got you a gift . . ." He looks grief-stricken.

"Well, I'll take an apple for the road. If you insist," you say, grinning.

You pluck an apple from the display and cheekily stash it in your bag. Then the smile fades from your lips. There are no more jokes to fill the gap, to soothe. Ian's hand is trembling slightly. His eyes.

"Beautiful," you say. You pronounce this biut-i-ful, framing his eyes with the two Ls of your thumb and forefinger. You will remember this. Around you and him, shoppers and tourists pass by obliviously, their laughter filling the long road, drenched in sunlight.

A LOVE STORY FROM THE END
OF THE WORLD

This will be the last place I look for you, Bada thought as the islet came into view. Her team's research vessel *Hope* was now almost six hours east northeast of Kvitøya, or 122 kilometers as the crow flies. They had made the journey on calm, lapis-blue waters free of ice floes. One hoped for more sea ice in April, but no one spoke aloud the obvious. As the ship drew nearer, the islet revealed its black-brown cliffs crested in white. Bada's spirits rose upon that sight. Somehow, the existence of the yearned-for ice cap made her think that she'd arrived at the final chapter of her search. *I'm at the end of the world, Umma,* she thought. *After this, I won't try to find you anymore.*

They spent another hour negotiating with the cliffs. At last, the ship crawled by the rocky beach on the opposite side and dispensed with the three scientists who would stay on the islet for the next six months. They hauled their tents, research equipment, satellite phone, and food up a bluff; a motorized lifeboat, in case of an emergency evacuation, was left in a cove. Once the move-in was finished, Bada and her colleagues ran over to the cliffs to watch *Hope* embark. It was to return every two months with supplies from the research station in Ny-ålesund. Everyone fell into silence when it glimmered one last time before disappearing over the horizon.

Bada's days began unfolding and rolling onto the next without pause. In the morning, she made tea on the hot plate and had a piece of toast. Then she went on a short walk to the cliffs with Hendrik, her Somali Norwegian colleague from the University of Oslo. Back at the camp, Bada washed up and got dressed for the day—these normal practices had to be maintained, especially when there didn't seem to be a point. At about 9 a.m., Hendrik, Ingrid, and Bada went down to the beach to collect seawater samples. They spent the afternoon analyzing data, reading other research papers, and writing their own. In the evening—or when the clock insisted that it was evening—they sat on camp chairs around their outdoor fire pit, eating freeze-dried meals and drinking hot cocoa, if they were feeling especially frisky. All night, the sun glided over the horizon like a seabird skating on the water's surface. Once everyone had tired of chatting or reading, Bada retreated to her tent and tried to sleep for a few hours without waking up from the light. Then her alarm would sound, and Hendrik would call out to her for their morning walk, and the routine would begin all over again.

Hendrik and Bada were collaborating on a study on the increased blue carbon sink potential of primary producers in the Arctic. In other words, climate change was warming the ocean and causing carbon-sequestering phytoplankton to flourish. Ingrid, also from the University of Oslo, was calculating the caloric and nutritional potential of the surging microalgae. Like the name of the ship that had brought them here, each of them had chosen to focus on the positive; to believe that in every situation, no matter how dire, *something* could and had to be done.

"That's what my mother told me: do something. Even if that something isn't enough to solve the world's problems, be on the side of good," Ingrid said one night around the fire pit. Tallest out of the three, she always wore a sweater and overalls, and her blond hair in two thick braids over her shoulders.

"Oh did she?" Hendrik said with a dark smile. Ingrid frowned, and Bada laid a hand on Hendrik's arm, but he swatted her away. "Was that why you were too scared to do anything when we passed by the dispatch ship?"

On their northward journey, only an hour away from the islet, *Hope* had passed by an oil and gas dispatch ship that would survey the Barents Sea for new offshore drilling. Hendrik had wanted to do something in protest—chant and pull up close to the ship, throw tomatoes at it, make a video and post it all over social media. Bada said she would help as long as Hendrik took the lead, but Ingrid had been furiously opposed, and they'd just sailed away.

"Stop, Hendrik," Bada said. But Ingrid raised a hand as if saying, *I can fight my own battles.*

"For the record, I wasn't the only one who wasn't keen on your plan, Hendrik," Ingrid said. "The captain wasn't sold and his word is the law at sea. And even if he wanted to, I don't feel comfortable breaking Norwegian law and vandalizing someone else's ship."

"You seem very fond of the law, Ingrid. Tell me then how Norway is *not* breaking the Paris Accord by producing even more fossil fuels. All of which it exports, of course, because heaven forbid that precious Norway is polluted by carbon emissions!" Hendrik rolled his eyes; Ingrid started to say something, but he continued over her protest.

"I'm not mad at you, Ingrid. But whenever I hear older Norwegians wax poetic about their sustainable coffee beans or some ancient grain milled using solar power, or a Michelin-starred chef plucking sea moss for a two-thousand-krone meal, I want to scream at those hypocrites."

"You shouldn't conflate me or even the older generation with the Norwegian government and powers that be. That's not fair." Ingrid's voice was shaking. "What about you? Isn't it hypocritical for you to be criticizing Norwegians as if you're not also a citizen? Or would you like to renounce your citizenship and go back to—"

"Oh my god, that's enough!" The shriek didn't come from Hendrik, but from Bada. It stunned all three of them into silence. Ingrid was the first to stomp off to her tent; then Hendrik left without another word. Bada waited for the fire to die and then went on a walk to the cliffs. She narrowed her eyes at the horizon, as if her mother would appear any moment.

The world looked like it was dreamed up by a child who only had orange paint. The sky, the sea, the ice covering the cliffs were all drenched in coral from the midnight sun.

When they were going through their orientation in Oslo, Hendrik had held out his cocoa-colored hand and introduced himself to Bada as "Hendrik Johansen."

"I'm actually not Norwegian, by the way," he said jokingly as they shook hands. "Despite my clearly Nordic features."

It turned out that Hendrik escaped war-torn Somalia in 2011 and was adopted by a Norwegian couple. He was grateful to them, mostly; he was already sixteen by that point, an age when a universally pitiable "refugee child" teeters into a frightening "male refugee." By a stroke of luck—or because he was constantly malnourished from birth—Hendrik was small for his age and could pass for a thirteen-year-old, which probably helped him to be adopted. The Johansens gave him their name and provided him with food, housing, and stability. Their well-intentioned interest was occasionally an adequate substitution for love. To this day, they had never once told him "I love you, Hendrik."

"It's not that they feel it but just can't say it. They can't say it because they don't feel it," Hendrik said without bitterness. "All the love I received and draw from, it's what I got from my real parents," he told Bada. Being an adoptee herself, she knew not to ask what happened to them.

Unlike Hendrick, Bada didn't have any memories of her birth parents and, in fact, didn't have any information about them at all. She was raised as Jeannette Mallory in Moscow, Idaho, by her white family—a minister, his stay-at-home wife, and their four children. Growing up, the only thing she knew about her origin was that she was adopted from an orphanage near Seoul in 1982.

The Mallorys weren't bad, as far as adoptive families go; there was no abuse of any kind. The adoptive mother hugged and kissed her just as much as her birth children. The minister, who was more emotionally distant, nonetheless made a sincere effort to talk to her about God once or twice a month. But the whole family was so conscious of not treating her any differently, that she couldn't help but feel that they always thought her as distinct. This was beyond the fact that she physically stood out from not only the family, but also the rest of Moscow and Idaho itself. But the Mallorys didn't like to hear her complaining about townspeople mistreating her. For example, when the old family dog was sick and the entire Mallory clan had gone to the vet, Jeannette had shown up late from softball practice. She asked if her family were in the examination room, and the receptionist wouldn't even tell her if they were there or not, saying "we can't reveal patient information." She left the waiting room and walked home alone, certain that if it had been any of the other Mallory children, they would have been allowed in. Things like this happened all the time in Jeannette's childhood. When she tried to talk about it with the family, their eyes turned glassy and their tone by turns doubting, defensive, and cold.

As soon as she turned eighteen, Jeannette begged her adoptive mother for the document from the adoption agency. It was written in Korean. At the first opportunity—a study abroad program in her sophomore year in college—Jeannette found her way to Seoul. Once in Korea, she thought she could find her birth mother. But it turned out that the document only had her birth name and the address of the orphanage, not the name or whereabouts of her mother. Undeterred, she sought out the adoption agency, only to be told that they couldn't reveal client information.

"Here's the relinquishment contract that she signed, which stipulates the birth mother shall remain anonymous in perpetuity," the receptionist explained, mixing English legalese into Korean. She discreetly checked the status of her polished nails while Jeannette pored over the waiver, also written in Korean. It was the original document, softened like a cotton rag through the years. Jeannette traced the thumbprint at the bottom in red ink, feeling for her mother's hand.

The receptionist said, "No touching, please. You can take a photograph, if you need."

"Let me just ask one thing. My name on the adoption agreement, Bada Moon, is that something the orphanage gave me? Or was that from my birth mother?" Jeannette asked desperately. "I've traveled halfway around the world to be here. Please."

The receptionist stopped admiring her nails and really studied Jeannette's face for the first time. She said, "Wait here," and then disappeared into a back room. She came out twenty minutes later, holding a yellowed piece of paper.

"I have an in-take note from the orphanage. Can you read Korean?"

"Only a little," Jeannette said. "I speak and understand better than I can read."

The receptionist unfolded the paper and started reading it herself.

"Name: Bada Moon. Sex: female. Birth date: 1981/12/15. Weight: 3.7 kg. Other: Appears to be healthy and well-fed. Brought by mother, bundled in a blanket with a stuffed whale. Said the whale calmed baby when crying. [Note: True that baby colicky unless given whale toy.] Showed reservation. Reassured best decision for child. Signed rel. contract on-site."

Jeannette didn't try to stop the tears running down her cheeks. She would allow herself that kindness. The receptionist handed her a tissue without saying anything else.

"You don't still have the stuffed whale? Or even just a picture . . ." Hope. Always rising, despite all.

"I'm so sorry," the receptionist said, not unkindly. "That's truly all we have."

Bada means "the sea" in Korean. *Moon* means "door" in Korean, but also "the moon" in English. She thought no mother would give a name like that to a child she didn't love. And the fact that she was brought in with the whale toy meant it really was her mother who named her, not the orphanage. She changed her name and became—or returned to—Bada Moon. The Mallorys didn't take well to this: her adoptive mother was hurt, and the pastor was offended. Why not just become Bada Mallory, if she were so keen on reconnecting to her heritage? Then a few years later, Bada came out, and that was the end of anything like normal family relations with the Mallory clan. For the past fifteen years, all Bada had to go by was an annual Christmas card.

She changed her major to marine biology and became an oceanographer. Everywhere she went, she looked for her mother. Seoul—where she returned many times, each time falling in love and becoming more certain of who she was. South Africa, France, Indonesia. She didn't think realistically that her mother would be outside of Korea, but it felt like a search nonetheless. And now she was at the end of the world, an unnamed islet so small as to not even exist on the map. After this project, she was going to interview for an assistant professorship at a few universities along the Eastern seaboard. She couldn't be a postdoctoral research fellow forever, and the forward movement in her career now was to put down some roots. That scared her more than being adrift her entire life, so she figured that was exactly what she needed to do. Before that, though, she wanted to feel her mother's presence at sea one last time.

By May, Bada arrived at the expected low point of any research project when the initial excitement died down and everything showed the chink of an impending doom. It started when another team from the University of Sheffield published a study in *Nature* about almost the exact same blue-carbon-sink topic, prompting Hendrik to yell, "Fuck those bastards!" They could avoid being redundant only if they pivoted their current project with the skill of an Olympic skeleton racer. And rewrite about ten thousand words of their thesis. Then, the ice cap was melting much, much faster than in previous years. Each day there was less white covering the cliffs, revealing bare rocks and even some mosses and lichen. Bada hardly needed more than a sweater or a windbreaker while collecting samples, and she feared what it would be like come September, when the ice was the lowest all year.

These trials and tribulations of fieldwork were always expected and usually overcome by banding together with colleagues. But Ingrid and Hendrik avoided each other as much as humanly possible, and Bada alone managed to keep some sort of camaraderie going. She eventually tired of acting chipper to bridge her two colleagues, and took to going on walks alone whenever she could.

It was on one of those solitary walks that Bada saw something strange in the sea. A bright white spot that wasn't just a cresting wave or clotted sea foam. It was moving with too much intention and speed. She ran to the edge of the cliffs as she realized that it had to be a polar bear, headed toward the islet. Bada saw two smaller clumps of white following the bigger clump: cubs. The mother bear soon reached the cliffs and heaved herself up on the lowest level of rocks, paying no mind to the little cubs struggling to climb. The first cub clawed its way up, and as Bada watched with her heart in her throat, the second cub finally managed to get a foothold. "Yes, you can do it!" she whispered.

The islet was hundreds of kilometers away from the nearest pack ice. This dense sheet of frozen sea was where polar bears normally hunted for seals—and created their dens for giving birth. These cubs would have come out into the outside world only a few months ago, but their home and hunting ground were all melting and they had to be on the move. They would have swum hundreds of kilometers and several days without stopping to get here. The mother bear looked tired, but still showed she'd be willing and able to keep going. It was the cubs that Bada worried about: the first cub was a bit bigger and following its mother closely, but the second one was clearly the runt and falling behind. It didn't have enough strength to climb higher and cried out to its mother in a high-pitched whine. The mother bear ignored the panicked call and scaled the wall toward her object, a dead ivory gull near the top of the cliff. It was a paltry prize after traveling all this way, but it would have to do. Once she scarfed down most of it, she let the first cub have a few last bites; but the runt, which had finally caught up to its family, had nothing left to eat.

The mother bear sniffed around the air, and Bada's blood turned cold thinking of the camp. She hoped that Ingrid and Hendrik had remembered to wash out their bowls and close their waste bin properly. It appeared that they had—the bear turned around and started making her way back down the cliffside. The first cub dutifully followed. But the second cub was now even slower, squealing every time it struggled to find a lower foothold. It clung to the ledge with its claws and dropped down to the next level, crying out for its mother, who was already in the water. Its sibling was now diving in. The runt was halfway down the cliff when it seemed to have reached the end of its strength. It stopped trying to follow its family and stood on all fours, crying toward the sea. Either unaware or indifferent to its calls, the mother bear and the bigger cub gradually became smaller amid the waves.

Bada ran to the camp. Hendrik was reading some papers and Ingrid was writing an email to her principal investigator. Bada explained the situation in a few sentences, talking to both colleagues together for the first time in weeks. Hendrik immediately said, "Let's go!" But Ingrid held up a hand and said, "Wait. I know you're trying to help. But I don't need to explain to you why it's harmful to feed a wild animal. You shouldn't interfere with nature in the name of conservation."

"Ingrid, literally *everything* humans do is interfering with nature. We'd be interfering even if we didn't feed this cub," Bada snapped. Ingrid crossed her arms and shook her head.

"Really, Bada, I expected more from you," she said. "I thought you were a *better scientist* than this." Then she stormed off into her tent. Hendrik and Bada looked at each other and shrugged, stopping just short of smiling.

"More cuteness for us, I guess," Hendrik said as they grabbed supplies.

When they reached the cliffs, the cub was standing immobile in the same niche. It had stopped crying, perhaps due to exhaustion. They decided to lower open cans of sardines using a jury-rigged basket-pole contraption. The "basket" was made of Ingrid's tote bag, which made them laugh. They dangled the contraption over the edge and peered down.

"This pole isn't long enough," Bada said in frustration.

"Wait. I think it's going to smell the food. Watch."

Like Hendrik said, the cub seemed to perk up upon smelling the sardines. It whined, pawing at the basket dangling above its head. Then it poked its claws into the ledge over its head. Swung its bottom side to side, and managed to stick its bottom claws on the ledge. With one last frantic push, it was on the landing with the sardines.

As the cub slurped down the fish, Bada felt that she had never done anything more right in her entire life.

If being more absorbed by caring for the cub than working on her research made her a "bad scientist," Bada supposed that Ingrid was right. She came to believe that protecting this one cub was more meaningful than any climate research that was read by a hundred rival scientists and maybe a few dozen journalists. What had that accomplished, compared to saving one real and sentient life?

As June warmth deepened into July heat, she occupied herself with feeding the cub out of their meager resources. *Hope* wasn't dropping off supplies until August. Hendrik and Bada pooled their resources together, but they had only so much powdered milk and cans of fish between them. When the arctic terns returned for their summer nesting, Bada took the cub out on a field trip and taught it to steal eggs. This way, she felt like she was teaching it a life skill to survive on its own—but truthfully, she knew it was a sham. Polar bears had to learn to hunt seals on ice. Not how to steal eggs on land.

Nonetheless, the cub—whom she named Hayang, meaning "white" in Korean—seemed to thrive. Without knowing much about polar bear anatomy, Bada decided that Hayang must be female, judging by her petite size. Hayang was now about a hundred pounds, a healthy weight for a cub in its first summer. She greeted Bada on her way back from the beach, jumping and making circles around her. When hugged by Bada, Hayang squealed joyfully in what sounded so much like laughter. Bada lay on the tundra and let Hayang rest her fluffy head on her stomach. Out of principle, Ingrid stayed out of these cuddle sessions, just as she didn't contribute to their "feed-a-bear" fund. But Hendrik almost always joined them. Some bright evenings when they lay intertwined together under the ceaseless twilight, Bada couldn't help thinking just one word, *family*, over and over and over again.

Then one morning, Bada was awakened by a loud BOOM. At first, she thought it must have been a dream. When she tried to go back to sleep, another BOOM sounded. By the time she straggled out of her tent, she realized the noise was striking every eight seconds from somewhere outside the islet. From the sea.

"Are you hearing this, Hendrik?" Bada called out. Hayang crawled to her side and whimpered, as if frightened.

"It's those fuckers on the fucking dispatch ship," Hendrick

said, already seated in his camping chair. "Seismic air gun to test for oil deposits."

"Uh, it's really loud," Bada said. She put her arms around Hayang and squeezed, and the cub relaxed in her embrace.

"Yeah, no shit. Louder if you're a whale."

"I *just* got used to the midnight sun. Not sure how I can sleep through this."

"You'll be awake for a while. They're going to be testing up and down these waters for three or four months. But who cares about sleep? In the grand scheme of things, none of our research matters when they're *still* pumping out more oil," Hendrik said bitterly. "I guess we just have to be on the side of good!" he shouted sarcastically in the direction of Ingrid's tent, which was extra silent as if to balance the cacophony outside.

A few weeks passed in which the morale at the camp fell to an all-time low. Ingrid now refused to speak to anyone, including Bada. Everyone else felt on edge from the constant blasts. It was there when they brushed their teeth; collected samples; tried to write their papers; ate a mug of oatmeal. Even Hayang lost some of her playfulness. Bada waited for *Hope* to bring supplies (more fish for Hayang) and news of the outside world beyond what they could gather from the weak satellite wi-fi.

One early morning in August, a week before *Hope* was due to arrive, Hendrik pulled Bada aside.

"I've got to do something, Bada," he said. "Even if you don't approve or help, I'm going to do it anyway. I'm just letting you know in advance because—"

"Because we're friends," she said. He smiled.

Hendrik's plan was this. He would take the motorized lifeboat to the dispatch ship. Finding it would be the easy part; it was announcing its location loudly every eight seconds. He would spray-paint "Oil drilling kills" on the side of the ship. Make a video of it and post it on social media. He had friends in NGOs who did guerrilla activism like this, they'd help his stunt get coverage by some journalists.

Bada listened in silence, and Hendrik interpreted this to mean she opposed the plan.

"Do you know what happened to my real parents, Bada?" he said. "We were supposed to get on the same boat, crossing the Mediterranean to reach Italy. The smugglers packed it full of hundreds of people desperate to leave Africa, and it capsized. There were plenty of ships nearby that could have come to their rescue. But no one responded to the Mayday call. So most of the people drowned. The only reason I survived was that I was pushed by the crowd, separated from my family, and prevented from getting on board."

"I'm so sorry," Bada said. Hendrik shook his head.

"My point is that I know what danger is, better than most people do. I don't take any of this lightly. But doing nothing is not an option."

"You'd have to take the satellite phone, in case of an emergency," Bada said. "There won't be any wi-fi here until you bring it back. Ingrid will be pissed."

"I wish there were two satellite phones so I could take both and make her even more angry," Hendrik said, smirking.

"Also, you might not have time to write 'Oil drilling kills.' How about just 'Oil kills'?"

"Genius is simplicity! I love it. Thank you." Hendrik smiled, and Bada bowed a little.

"But seriously, Bada, I guess this means you're not coming?"

Bada sighed. She realized how much she liked Hendrik, this

Somali guy she'd met only four months ago, who came from a vastly different background, who looked and moved differently through the world. Yet she felt a kinship to him that erased all those distinctions.

"I wish I could," Bada said lamely. Hendrik made a sign with his hand as if to say, that's enough.

"No pressure. Just help me get the supplies to the beach."

Within an hour, Bada and Hendrik pushed the lifeboat together into the water. Hendrik jumped into the boat and paddled away from the shore.

When he had safely made it to deeper waters where he could start the engine, Bada shouted, "I love you, Hendrik!" He turned around and waved at her. And then he was off.

It took Ingrid some time to realize that Hendrik was gone because she was so used to ignoring and avoiding him. But in the afternoon, she realized the satellite phone was missing and truly blew up at Bada for the first time.

"You let him go alone into the middle of the Barents Sea, to do some weird eco-terrorism against an oil ship?"

"It's a nonviolent protest. Hardly eco-terrorism," Bada countered.

"Do you know these ships have commandos? They carry guns, Bada. They could shoot him."

Bada hadn't known that, but she had a feeling it wouldn't have mattered to Hendrik anyway. Ingrid didn't care about Bada's excuses. According to her, the worst offense that Bada committed was that she let Hendrik abscond with their only means of communication with the outside world. They couldn't reach Hendrik or the Ny-ålesund research station. Bada pointed out that *Hope* was due to arrive in just a week, and Ingrid seemed to decide that the silent treatment was the best course of action.

For the first twenty-four hours, Bada listened to the continuous seismic blasts, thinking that if Hendrik reached the ship (which couldn't be more than an hour away from the islet), there would have been an interruption. It surprised and dismayed her when no such pause happened. Had Hendrik gotten lost at sea? She didn't like to think of what might have happened if he really ran into armed commandos.

Adding to her despair was the fact that she was completely out of food for Hayang. She herself was subsisting on only oatmeal with water and some tasteless ramen noodles, but she felt guilty when she was eating these while Hayang looked on with hungry eyes. If polar bears could live on her food, she would have gladly given it to Hayang and starved herself until *Hope* arrived. The only thing she could do was try to look for dead birds and unguarded eggs at the arctic tern colony.

Polar bear mothers have one of the longest fasting periods of any animal. For eight months around pregnancy, birth, and nursing, polar bear mothers stay in their pack ice dens without eating or drinking. Such maternal duress is only rivaled by some species of whales. Like other animals whose parents make extraordinary sacrifices for their offspring, polar bear mothers form a strong bond with their cubs. But some mothers abandon their young when they are faced with severe lack of resources. Yet Bada doesn't think this explains everything. Why, then, do more bear mothers reject their young in captivity, where they don't struggle against hunger or other dangers?

At last, there was a ship on the horizon. It wasn't Hendrik's lifeboat or *Hope*. As it gradually approached the islet, Bada realized it was a ship of the Norwegian coast guard.

The officers waved when they stepped off their dinghy and

onto the beach. Both Ingrid and Bada were there, collecting sea-water samples. Bada was for some reason glad that they were not at the camp, where Hayang might make an appearance at any time. One of the officers, who looked like the one in charge, asked for their names and nodded as they introduced themselves.

"Your colleague, Hendrik Johansen, was arrested three days ago for attempted vandalism and trespassing on an Equinor property," the officer said. "He's being transferred to a jail on the mainland."

"Is he okay?" Bada said, and the officer shrugged.

"He isn't injured, if that's what you mean. He'll need to find a good immigration lawyer, if you ask me."

"He's a Norwegian citizen," said a trenchant voice, and Bada turned around in surprise. Ingrid had broken her silence to stand up for her longtime nemesis.

"Eh, yes and no, right?" the officer muttered, but Ingrid didn't budge.

"Hendrik Johansen is a naturalized citizen of Norway. You can't extradite him," she said, glowering.

"If you say so," the officer pushed back. If he was more neu-trally disposed before, he was now definitely annoyed. "At any rate, Equinor is arguing that this islet is also included in the feder-ally approved exploration site, and therefore you're all trespassing. You need to pack up and leave. Immediately."

Ingrid, who had been speaking to the officer in English, switched to rapid-fire Norwegian—as did the officer, leaving Bada in the dark. After much heated back-and-forth, the officers retreated to their dinghy and pushed off from the shore.

"What happened?" Bada asked, watching the officers clamber onto their ship.

"I argued that we'd received permission through our universi-ties to be exactly where we are, it was cleared with all governmen-tal parties thereof, and that we weren't going anywhere with them as if we were criminals," Ingrid said. "But he said we'll have to return to Ny-ålesund on the next *Hope* journey and await further bureaucratic bullshit."

"You think they'll allow us to come back?"

Ingrid shrugged. "Who knows? But I wouldn't count on it." And then she really surprised Bada by crouching down on the ground and burying her face in her arms. Bada rubbed her back, letting her cry in peace. Their respective research projects were completely destroyed; not just the entire spring and summer fieldwork was worthless, but (for Ingrid) her PhD was now derailed.

The ramifications for Bada's career must also have been dire, but she didn't care about any of that. Suddenly, it dawned on her that she would have to leave Hayang alone in a matter of days. Of course, Bada was always supposed to leave in October, but for some reason that had felt far enough away. Now, she couldn't avoid the facts. In the wild, polar bear cubs stay with their mothers for two and a half years. Hayang was just eight months old—she had no chance of surviving on her own. Bada had kept her alive only to abandon her to her fate. But the other option was to bring her along so that she'd always be dependent on humans. Hayang would quickly outgrow the care that Bada could provide for her. But she wouldn't be able to return to the wild, either. Bear cubs like Hayang always ended up in a zoo—and that fate seemed to Bada much worse.

Bada decided to spend every minute of the last few days on the islet with Hayang. She took the cub to the beach and made fishing movements, striking at the water's surface. By some deranged reasoning, she thought that maybe Hayang could pick up how to hunt for fish or seals by watching her. Hayang seemed to think it was a game and splashed into the sea, peeking over her shoulder at Bada as if saying, *Look at me, Mommy!*

And when they tired of jumping in the sea, they combed the cliffs and stalked the birds. When they tired of that too, they rolled around on the tundra. The moss was springy and soft on their backs. *Look!* Bada pointed at a single white flower. *A white dryad.* It wasn't supposed to be up this north, was it? There wasn't even a bee to pollinate it for miles of open sea. But its seed had traveled here anyway. Decided to bloom where there was no other of its kind.

I love you, Hayang. Bada locked her arms around the cub and buried her face in her soft fur. Hayang, a little spent from the lack of food, but still playful and happy, rubbed her snout on Bada's head. In fact, Bada was hungry too. She wished she could be even hungrier so that Hayang wouldn't be the only one suffering.

The day *Hope* was due to arrive, Hayang seemed to understand that Bada was expecting something. She showed extraordinary patience and trust. They were waiting together on the cliffs, scanning the horizon.

Bada was remembering the first time she ever saw the sea, just before she changed her name. Until then, she'd only ever been to slime-bottomed lakes in landlocked Idaho. She took an early morning bus to Gangneung because her Korean lab mate had recommended it. She got off at the last stop and looked around for a place to get breakfast. She was hungry but too intimidated and poor to try a restaurant. There was a corner store with a couple of plastic tables and chairs outside. She bought a cup ramyun, and the *ajumma*-owner filled it with hot water.

The *ajumma*—a woman of indeterminate middle age—said to Bada, "You should have come during summer! Then you could have gone swimming. But truth be told, I love the winter sea the best." The *ajumma* quickly realized Bada spoke limited Korean and asked where she was from. "How are you so little? Coming from America?" The *ajumma* made an outline of Bada's petite body with her hands, to make sure her point got across. "You need to eat more. Your mother would worry," she said, clicking her tongue and handing her a Choco Pie, on the house.

To this day, Bada could recall no meal more clearly than that orange ramyun broth, its steam wafting in the frigid air, the woody taste of the disposable chopsticks, and the marshmallow cream inside the chocolate pie. Afterward, she made her way to the jetty, pulling down on the sleeves of her coat to hide her bare hands. The wind made her eyes water. But she didn't feel cold or alone: she was filled with only a sense of being exactly where she needed to be. When she stood on the edge of the East Sea, an epiphany, a profound recognition, had coursed through her as if she were a wind chime in God's garden.

Now, gazing at the Arctic with Hayang by her side, Bada felt as though she was having another epiphany—but in reverse. To love, then to abandon, and then to continue living was an unimaginable ringing of the spirit, a bronze bell struck again and again to oblivion. She softly called out to her mother.

Then Bada saw something moving in the waves. She sat up, her heart thundering in her ears. Hayang also noticed the same thing, her round ears perking up.

It was a white clump bobbing in the water, followed by a smaller white clump.

Hayang let out a soft, keening call.

The mother bear and the cub heaved themselves up onto the rocks. Sniffing around, the bear stood on her hind legs and looked up at the top, where Bada and Hayang were seated. Hayang cried out a little louder. The mother bear started scaling the cliffside, followed by her cub.

Hayang stood up, trying to get a better view. Then she started descending the wall with surprising speed. The bear and her other cub stood in place, waiting for Hayang to join them. Hardly were they reunited when they headed back to the sea, in the order in which they first arrived at the islet. The mother bear dove in first, followed by the bigger cub. And then just before the ledge, Hayang turned to look back at Bada, who was sobbing despite herself. Hayang whined softly.

Go, don't lose them! The cub snorted as if she understood Bada and splashed into the azure waters after her mother and sibling. *I love you until the end of time, Hayang,* Bada said with all her heart. And also, *Thank you for bringing them back, Umma.*

And Bada lay there watching until the three white dots could no longer be distinguished from the waves, the sea, the sky—everything that made up this hard, ugly, broken, and beautiful world where, in the end, absolutely nothing mattered except how much one loved.

ACKNOWLEDGMENTS

By the time this book is published, my agent, Jody Kahn, and I will have been working together for exactly ten years. That November 2015 evening when she agreed to represent me remains one of the happiest moments of my life. It changed everything for me. We've published three books together in a decade, which is even more un-likely considering how it all began. Thank you, Jody, for being the first person to believe that I am a writer, and for championing me with your incredible grace and intellect. I also thank everyone at Brandt & Hochman, the classiest literary agency in New York.

This book would not have been possible without Deborah Ghim's brilliant and generous editing. Thank you, Deborah, for supporting my vision while asking it to become the best it can be. Thank you Shelly Perron for copyediting and Allison Saltzman for this evocative cover; Helen Atsma, Miriam Parker, Cordelia Calvert, Will Howard, Catherine Barbosa-Ross, and the rest of the Ecco team for all the seen and unseen efforts to bring this book to life.

I'm also indebted to Regional Arts and Culture Council for supporting the writing of "A Woman's Life, in 10 Scenes," which was first published in *Joyland*. Thank you also to editors of *Zyzzyva, Catapult, Dispatches from Anarres: Tales in Tribute to Ursula K. Le Guin, Brink*, and *Catamaran* for first publishing stories in this collection.

I feel sheepish writing two acknowledgments in as many years. In fact, I didn't write a book a year—this collection gathers stories I've written from 2015 to 2024, a decade in which I've been supported by so many people. I will mention just a few by name and hope that others will forgive the omission. Vivian Lee, a fellow Korean vegan environmentalist whose inexhaustible energy and sincerity have inspired me for years. Vladimir Ilyich Tolstoy and his daughter, Anastasia, for carrying on the legacy of Leo Tolstoy, whose art and humanitarianism have been my lifelong North Star. My readers from around the world—so many of you read *Beasts of a Little Land* and then, as soon as possible *City of Night Birds*, or vice versa, and told me you were waiting for yet another book. You probably won't ever know how many times your messages saved me from dejection. It's an immense privilege to be sharing my work with you, and I will never take it for granted. Thank you, Renee Serell, for going through life with me. I'm grateful to Elise Anderson, who is so much a part of my creative process that we were even, just ten minutes ago, discussing parts of this book and the last one. May we be each other's mirrors until the end.

My parents have instilled in me a lifelong love of nature and literature. Their potager garden, roses, fruit trees, poems, their mountain trails, and their sea are woven into this book, whether it's visible or invisible. This book is my father freeing a rabbit that was ensnared in his antirabbit fence, during the great Oregon wildfires of 2020; the back of him as he muttered wistfully, "I wish I could drive it somewhere far and drop it off," and the rabbit hopped away and disappeared into the safety of his vegetable patch. This book is my mother walking by the gray Pacific until she's a dot on the long, long beach, and her poem at an outdoor exhibit on the green lawn of the Korean Society of Oregon. Without them, I am nothing; I owe them everything.

Last, but not least, thank you David (and Ody and Kili) for your love. I fly home to you.

AUTHOR'S NOTE

What can literature do? *Que peut la littérature ?* asked the infamous 1964 debate at the Mutualité in Paris. On one side stood committed-literature figures like Sartre and Beauvoir, and on the other, Jean Ricardou and the *nouveaux romanciers* (New Novelists). Ricardou criticized Sartre for saying that *Nausea* is worthless in front of a dying child; to Ricardou, pitting the value of a human life against that of a book wasn't a fair comparison. He denounced both art for art's sake (*l'art pour l'art*) and art for humanity's sake (*l'art pour l'homme*), arguing that art *is* humanity itself.

While I don't see myself as wholly aligned with Ricardou, his words have stayed with me over the years. We don't create or appreciate art for its moral utility. (Propaganda novels, films, paintings, and sculptures all serve to remind us of this fact.) But insofar as art is humanity, shouldn't art make us more human, not less?

Is an artist's duty limited to her own art (her books, for example)?

Or does her duty also extend to art in general and to humanity in general?

I believe that yes, it does. My belief doesn't truly come from literary theory, but from my own experience of writing. We live in a world of artificial intelligence and aesthetic luxuries beyond calculation. Yet what we recognize as art is the humanity of the artist, which serves to remind the reader, the viewer, or the listener of their own humanity. So, what heals or endangers humanity should be the concern of every artist.

If you're an artist, it is not conscionable to use our ecological catastrophe as material for fiction and not personally do something to help. Over the past two decades, I've learned a lot about the ways you can reduce your ecological impact and save our Blue Planet. Here's the one thing I hope you take away: A seminal 2018 metastudy published in *Science*[1] revealed that the single most effective change you can make to slow the impacts of climate change is to go vegan. Led by Joseph Poore at the University of Oxford, the groundbreaking study concluded that an average vegan has a 73 percent smaller carbon footprint than an average eater. Even more remarkably, a 2023 study published in *Nature*[2] was able to corroborate the earlier results, thus bringing the world's top two scientific journals in agreement: vegan diet reduces emissions, water pollution, and land use by 75 percent; destruction of wildlife by 66 percent; and water use by 54 percent. We currently use half of the world's vegetative land for agriculture;[3] and with Earth's population on track to rise to ten billion by 2050, going plant-based is not a fanciful idea but a moral imperative necessary to feed all humanity. Likewise, a vegan diet is so effective for wildlife conservation because habitat destruction is the leading driver of extinction. In

---

1   Joseph Poore and T. Nemecek, "Reducing food's environmental impacts through producers and consumers," *Science* 360, no. 987–992 (2018), https://josephpoore .com/Science%20360%206392%20987%20-%20Accepted%20Manuscript.pdf.

2   Peter Scarborough et al., "Vegans, vegetarians, fish-eaters and meat-eaters in the UK show discrepant environmental impacts," *Nature Food* 4, no. 565–574 (2023), https://doi.org/10.1038/s43016-023-00795-w.

3   Weronika Strzyżyńska, "Can the world feed 8 bn people sustainably?" *The Guardian*, November 15, 2022, https://www.theguardian.com/global -development/2022/nov/15/can-the-world-feed-8bn-people-sustainably.

my experience, concern for wildlife is the second most compelling factor (after climate) for many people.

Personally, I became vegan in 2006 for livestock animals. I believe these sentient and intelligent beings deserve moral consideration just as much as wildlife species do. It is by far the best decision I've ever made in my life.

Around here is when someone usually offers a rebuttal that putting the onus on individual lifestyle choice takes away from holding politicians and corporations accountable. My answer is that you should do both; vegan diet and political activity are not mutually exclusive. We need to be voting and marching in nonviolent protests. But if you want to make an immediate difference to the planet (and a life-giving difference to individual animals), the world's best scientists suggest the most effective way is to become vegan. If that seems overwhelming, just intentionally adding more vegan meals into your week would also help tremendously. Most of my loved ones are not vegan, but many of them make conscious efforts to eat more plant-based; put together, these flexible eaters are arguably more effective in bringing a sea change than the few strict vegans like myself.

After I changed my diet, I also adopted other sustainable habits one by one: composting; cruelty-free personal and home products; secondhand or sustainable vegan clothing, furniture, and accessories; organic produce (for farmworkers' health); low-plastic consumption. You don't have to do everything perfectly—I definitely can't. Celebrate the effort, and you might find joy, beauty, or satisfaction in unlikely places.

If none of these suggestions resonate with you, find something that you truly feel passionate about and give this cause your sustained energy and time. Each of my three books has an associated cause to which I donate a portion of my proceeds. My first book was dedicated to Siberian tiger and Amur leopard conservation,[1] and my second to Somalian development and aid.[2] It's tiger and leopard conservation that taught me how many years it takes to even understand my comprehensive role and long-term goal within a cause, but this realization has been incredibly rewarding. For this book, I changed my strategy: instead of choosing charismatic megafauna or a global cause, I am giving microgrants to grassroots organizations that are making tangible and life-changing differences in their local communities. The first two of these are Neighborhood Cats[3] and Mina's Dream Space, both animal rescue nonprofits that are volunteer-run by some of the most selfless people I've ever met. Most of us already have connections to a nonprofit or interest in an issue, but the effort is often not sustained enough to make a lasting impact. We know that love isn't fleeting; likewise, fleeting compassion isn't compassion at all. If we all gave long-term committed effort to one or two chosen causes, I believe that the world would change. We have within ourselves everything we need to save humanity.

---

1   From 2020 until early 2023, I worked with Phoenix Fund, based in Vladivostok, Russia. After political circumstances made further collaboration impossible, I have been contributing to and working with the scientists of Korean Tiger Leopard Conservation Fund and the Land of the Leopard National Park. Tigers and leopards were extirpated from the Korean peninsula in large part due to war, and civilian scientists across borders are working to prevent these animals from becoming fully extinct because of present-day man-made conflicts. Find out more about KTLCF: blog.naver.com/savetiger.

2   Caritas Somalia is a branch of Caritas Internationalis, the charitable arm of the Catholic Church, and has been carrying on lifesaving work through decades of droughts, famines, and civil war. Find out more: caritas.org.

3   Find out more: neighborhoodcats.org.